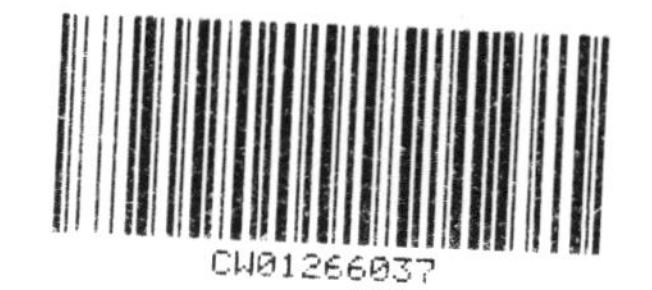

Microcontroller Know How

Amateur Radio projects and much more

By Mark Jones, G0MGX

Published by

Radio Society of Great Britain of 3 Abbey Court, Priory Business Park, Bedford MK44 3WH, United Kingdom
www.rsgb.org

First Printed 2021

ISBN: 9781 9139 9508 9

Cover design: Kevin Williams, M6CYB
Editing: Ed Durrant, DD5LP
Production: Mark Allgar, M1MPA
Typography and design: Chris Danby, G0DWV

Printed in Great Britain by Short Run Press of Exeter, Devon

Any amendments or updates to this book can be found at:
www.rsgb.org/booksextra

Contents

Foreword

Microcontroller units (MCUs) are simply everywhere you look in our modern lives. Just about every electronic gadget you interact with will contain one. From your electronic toothbrush, your television even your washing machine; they all contain MCUs and associated embedded software. If you have driven anywhere in your car then an MCU will have undoubtedly been controlling a number of the engine components while you travel. If you are the owner of a modern amateur radio transceiver then this unit will inevitably contain one or more MCUs to control displays, interface with your PC, configure the radio settings and do numerous other things. When you update the 'firmware' in your rig, you are updating the embedded software that runs on one or more of the internal MCUs.

After reading this title, you shouldn't expect an immediate career in embedded microcontroller programming, but you can expect to be able to tackle the projects contained in this book and also embark on some of your own. I also hope you will have a better understanding of what exactly a microcontroller is and how they work.

All the projects will be developed in a Microsoft Windows 10 environment and written in the software language "C". If you are a Linux only user, all the software we are going to use is also available for that platform.

I hope to take you on a journey that will teach you some of the basics of embedded programming using the "C" language and also show you a range of projects from the relatively simple to the more complex.

Please join me on this journey and see what we can learn as we go.

Listed files, suggested hardware purchase URLs and any updates to this book can be found at www.rsgb.org/booksextra

An Introduction to Microcontrollers

1

A microcontroller unit (MCU) is a generic term for a small computer contained in a single package. An MCU will contain a central processor unit (CPU) plus memory and other programmable input/output peripherals. MCUs are designed specifically for use with embedded applications and can operate stand alone, unlike the microprocessor in your home PC, which needs other external components to make a working system.

Different MCUs contain varied architectures and input/output peripherals depending on their target application. A system designer can select the MCU most appropriate for their needs, all MCUs contain a number of General Purpose Input Output (GPIO) pins which are software configurable. When configured as inputs they can be used to read signals from external sensors, when set to be outputs they can light LEDs, drive motors (normally indirectly via external power electronics), switch on external devices et cetera.

Many embedded applications need to read analogue inputs from external sensors, thus the majority of MCUs contain analogue to digital converters (ADCs) to facilitate such interfaces. Some MCUs contain digital to analogue converters (DACs) allowing variable voltage outputs to be produced.

Most, if not all MCUs also contain counter and timer functions. These counter/timers can be configured through software in many different ways. Common uses include counting external pulses, tracking elapsed time or prompting the embedded software that it's time to do something important. Some MCUs also contain a dedicated timer module called a Real-Time Clock (RTC). This is a specific timer block dedicated to accurate time keeping and can be used for timing applications including alarms.

A very generic illustration of a MCU architecture is shown in **Figure 1.1**.

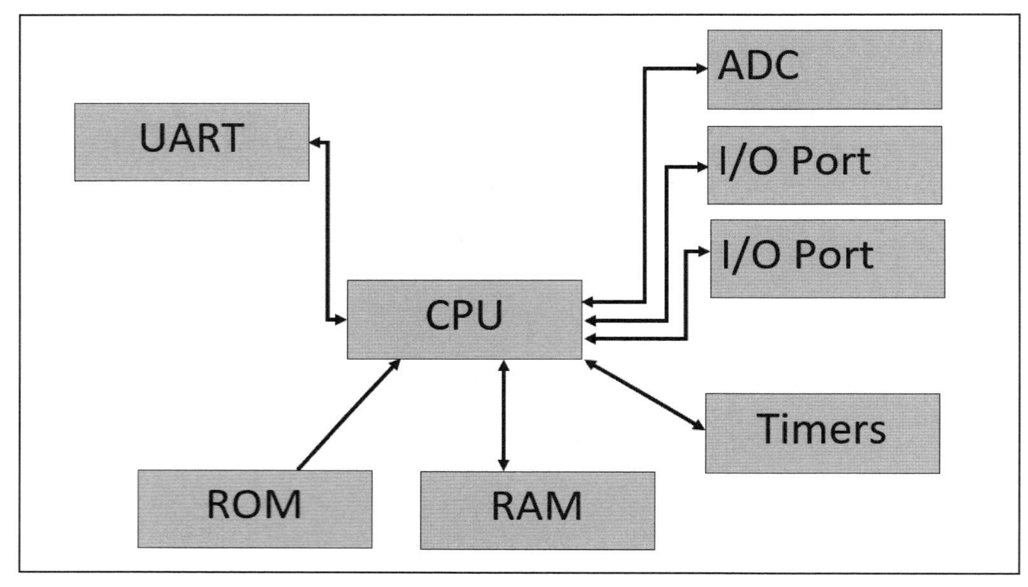

Figure 1.1 - Generic MCU Architecture

Other input/output peripherals commonly found in MCUs include:

- I2C – inter-integrated circuit (pronounced I 2 C) this is a simple serial data communication protocol used to communicate short distances between different integrated circuits. An example might be the communication between an MCU and a display driver IC.
- SPI – Serial Peripheral Interface – this is a simple synchronous serial communication protocol using a master slave architecture.
- UART – Universal Asynchronous Receiver-Transmitter – this is the serial communication protocol typically found in a PC's COM port.

Like all computers the MCU needs a system clock; this is the fundamental signal on which all actions are synchronised. Clock speeds vary as do their sources. An MCU will typically contain an internal clock source (often called the HSI – high speed internal) and will also have the interface capability to be driven from an external clock source (often called the HSE – high speed external). If the MCU contains a RTC then there is normally a secondary internal clock source (often called the LSI – low speed internal) and a facility for a second external clock (often called the LSE – low speed external).

Often the selected clock can be routed through an internal Phase Locked

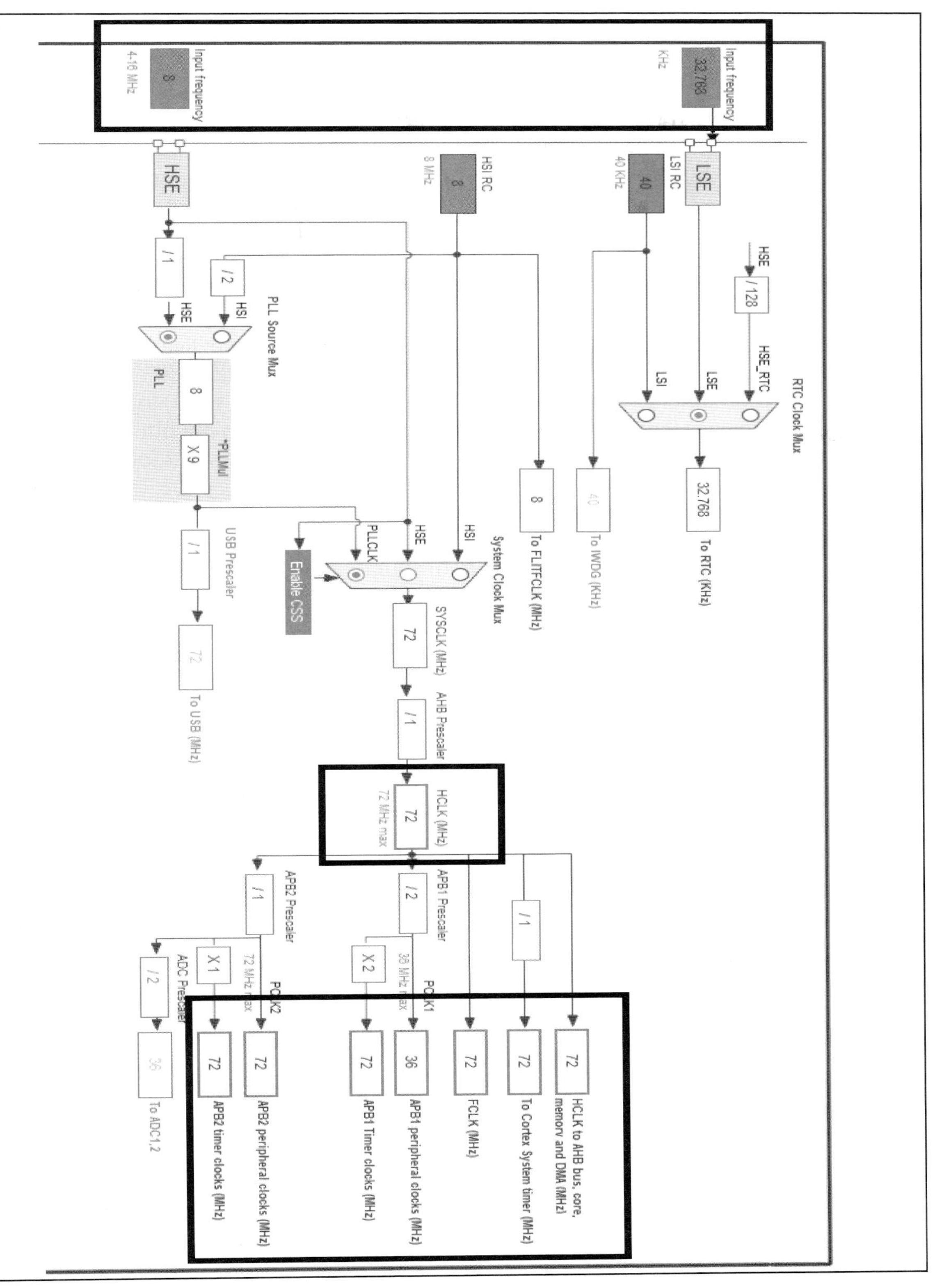

Figure1.2 - Example MCU Clock Architecture

Loop (PLL) which is used to alter the source clock speed. All this internal MCU configuration will be controlled by software during system initialisation.

In **Figure 1.2** the main external clock (HSE) on the left-hand side of the diagram is 8MHz. This is fed through various software configurable prescalers, switches and a PLL to generate the main clock frequency shown near the centre of the diagram (HCLK) of 72MHz. This means (very roughly) the MCU can execute an amazing 72 million instructions per second.

You will also see in Figure 1.2 that the RTC is configured to use an external low speed clock source (LSE) of 32.768 kHz.

There are many MCU manufacturers in a very competitive marketplace. Prices are varied but range from incredibly low-cost devices (there are one-shot programable MCU devices available for a few pence each) to moderately priced devices that can be individually purchased. I recently bought a 40 pin DIL packaged MCU and found that a good quality turned pin socket for the device was more expensive than the MCU itself!

We are now going to look at a few examples of MCUs and select an inexpensive, commonly available, and flexible variant for our projects.

A word of warning; please beware of counterfeit MCUs. From the land where copyright means "If we're going to copy it, let's copy it right" we find a wealth of counterfeit products in the marketplace. In my experience buying from a reputable distributor at slightly higher cost is often worthwhile in the long run.

STM32

2

Overview

STM32 is a family of 32-bit microcontrollers produced by STMicroelectronics. They are based on the Arm® Cortex®-M processor series. This means that each MCU comprises an Arm® Cortex®-M processor with additional peripherals added by STMicroelectronics into a single package.

The Arm® Cortex®-M processor cores are licenced by Arm Holdings and are 32-bit Reduced Instruction Set Computer (RISC). RISC processors have their origins as far back as Alan Turing's Automatic Computing Engine (ACE), but IBM were the first to market a processor as using a RISC architecture as early as 1980. The basic concept is that the processor uses a small number of clock cycles per instruction as each instruction is highly optimised for speed. As the name implies, there are a reduced number of available instructions compared to non-RISC processors. It is estimated that ARM processors are embedded in tens of billions of in-service devices worldwide; they are also hugely popular with developers.

A number of the Arm® Cortex®-M processor cores also include a Floating-Point Unit (FPU) which is a dedicated unit for floating point arithmetic. These cores are known as Cortex-Mx with FPU or Cortex-MxF. STM32 MCUs containing FPUs are prefixed STM32F.

The STM32 microcontroller family contains many different MCUs. Some are optimised for performance, some for mainstream use and others for low power consumption. The STMicroelectronics website contains a large amount of information on their different microcontrollers.

Development Boards

A large number of the MCUs manufactured by STMicroelectronics are incorporated into their own range of development boards known as "Nucleo" boards. These boards are priced very competitively and are intended to provide easy access to the MCU family for software developers. There are four basic types of Nucleo:

1. Nucleo-32; These are small footprint boards (the same size as an Arduino Nano)
2. Nucleo-64; Medium sized boards with more GPIO pins accessible than the Nucleo-32
3. Nucleo-144; Large boards with all interfacing pins accessible
4. Discovery Kits; these are MCU based boards with additional peripherals like touch screens and media interfaces.

For our initial project we will use an extremely popular and inexpensive development board known as the "blue pill". This board contains:

- STM32F103C8T6 MCU
- 20kB RAM
- 128kB FLASH (ROM)
- 8MHz HSE crystal oscillator
- 32.768 kHz LSE crystal oscillator
- 3V3 voltage regulator
- USB connection

This board can be programmed using the Arduino IDE software.

Figure 2.1 - Some of the Authors STM32 Development Boards

Figure 2.2 shows two examples of development boards with the STM32F103 MCU. The board on the left is the STMicroelectronics Nucleo-64-F103RB and the board on the right is the "Blue Pill" we will be using.

Figure 2.2 - Authors STM32F103 boards

Arduino Installation and Configuration

We will now go through the process of downloading the Arduino IDE and configuring it to work with our STM32 board.

Shopping List

This section of the book uses the following components (Please refer to the hardware list file on the Books extra page for this book to obtain up-to-date URLs for where to buy these items):

- STM32F103 Blue Pill development board
- FTDI USB to Serial adaptor
- Breadboard and jumper wires
- USB cable to suit first two items above

Step 1 – IDE Installation

If you have used the Arduino IDE previously then great! However, if you haven't, I would like to do all I can to encourage you to dive in and have a go. Software development is really nothing to be afraid of and I hope we can tackle the steps together.

We first need to go to the Arduino home page:

https://www.arduino.cc/ and follow the links to download the IDE.

At the time of writing this is found in Software, Downloads and then under a section titled "Download the Arduino IDE". You need to download and install the software; once the installation is complete you can run the software for the first time. You should see the same as **Figure 2.3**.

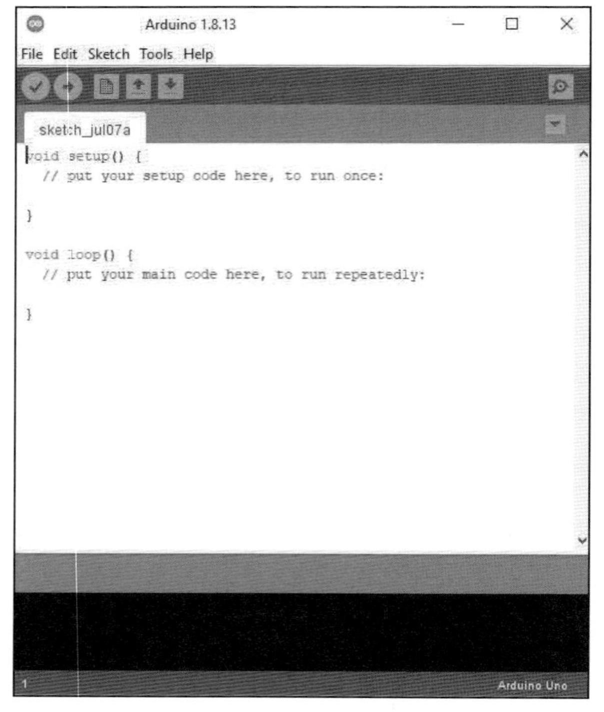

Figure 2.3 - Arduino IDE on first execution

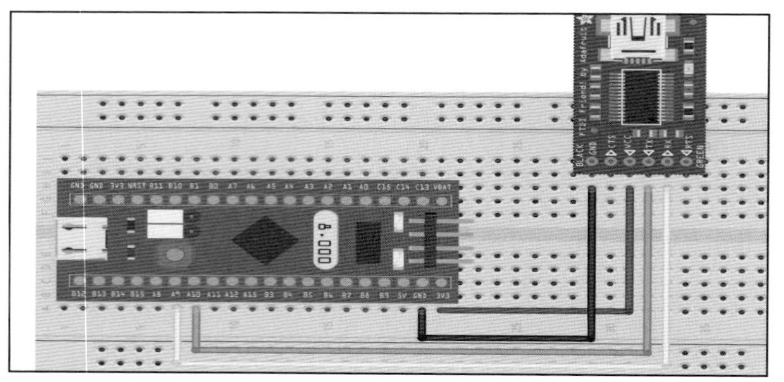

Figure 2.4 - Bootloader load configuration

STM32 Blue Pill	FTDI USB to Serial Adaptor
3V3	VCC
G (Ground)	GND
A9	RX
A10	TX

Table 1 - Blue Pill to FTDI Interface Connections

Step 2 – Arduino Bootloader Installation

There are a number of options for communication between the Blue Pill board and the Arduino IDE, for our projects we will install a bootloader to take care of this for us onto the blue pill board using a FTDI USB to Serial interface board.

Configure your items from this chapters shopping list in accordance with **Figure 2.4** below:

Don't worry if your FTDI USB to Serial adaptor board is different to the illustration. We need to connect the pins in accordance with **Table 1.**

It is **very important** that you set the jumper on the FTDI USB to Serial adaptor to the 3V3 setting. The STM32 is a 3V3 device and the logic levels and power supply need to be at this voltage.

Now connect a USB cable between the FTDI USB to Serial adaptor and your PC (you did check the jumper is set to 3V3 before doing this, didn't you?). Windows 10 should automatically install the appropriate driver and a new COM port will appear in Device Manager.

To access device manager, type "Device Manager" into the windows 10 search box and then select the application listed, or alternatively run the Device Manager.bat file included within the "scripts" directory of this books extra page url.

If you find your new device looks like **Figure 2.5** then right click on the device in Device Manager and se-

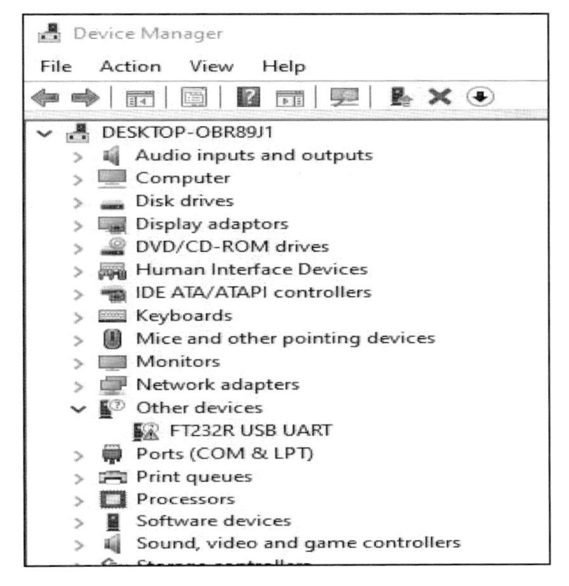

Figure 2.5 - Device Driver Error

Figure 2.6 - COM Port Driver Installed Correctly

lect Update Driver. Once this action is complete you should have a COM port as shown in **Figure 2.6**.

Once installed correctly, note the COM port number that is assigned to your FTDI USB to Serial adaptor, we will need this later in the process.

Step 3 – Flash Programmer Installation

We now need to download and install the STM32 Flash Programmer application provided by STMicroelectronics. At the time of writing this was available at:

https://www.st.com/en/development-tools/flasher-stm32.html

The software is also included in the downloads directory of this book's Books extra page in a folder called STM32 Flash Loader.

Once you have the file, we need to install this utility, so run the file and follow the instructions.

Step 4 – Bootloader Download

We need to download the actual bootloader software that we are going to install on our STM32 board.

At the time of writing this was available at: https://github.com/rogerclarkmelbourne/STM32duino-bootloader/tree/master/binaries and the file we need is listed as: generic_boot20_pc13.bin

The file is also contained in this book's extra page url in the STM32 Bootloader directory.

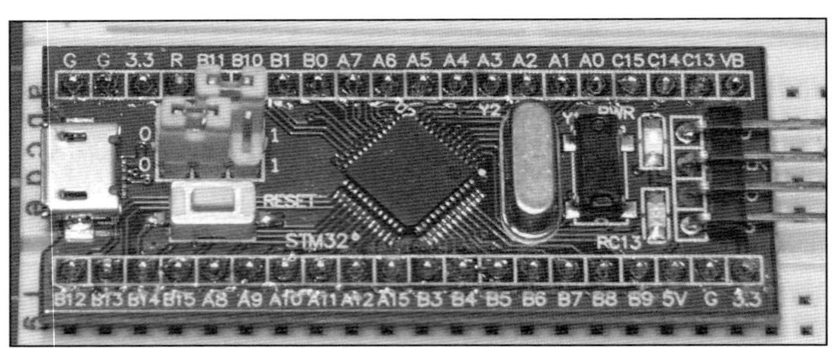

Figure 2.7 - Boot 1 Enable

Step 5 – Flashing the Bootloader

To program the bootloader into the STM32 Blue Pill board, make sure the USB cable from the FTDI USB to Serial adaptor is connected and the BP board has power attached. Move the jumper on the STM32 Blue Pill board marked Boot 0 to the 1 position as shown in **Figure 2.7** and press the reset button.

Run the Flash Programmer we installed at step 3 and make sure that the COM port selected matches the number we noted down in step 2. An example is shown in **Figure 2.8**.

Once you have the correct port selected, click on Next to advance through the program settings.

The Bootloader programmer should now confirm that the target is accessible as shown in **Figure 2.9** and also confirm the FLASH memory size of your board.

If you then click next, the FLASH Bootloader program will describe the type of board as "Medium Density" as shown in **Figure 2.10** .

Click Next once more and you should now be able to select the option "download to device" and then select a file type of .bin (by default this is set to .s19) and then select the binary file we downloaded in step 4 above. This is shown in **Figure 2.11**.

Click next for the final time and the FLASH programmer should now program the bootloader into the device.

Once complete you should see the programming software display look like **Figure 2.12**.

It is now essential that you put the boot 0 jumper back to the 0 position prior to pressing

Flash Loader Demonstrator
life.augmented
Select the communication port and set settings, then click next to open connection.
Common for all families
UART
Port Name COM5 Parity Even
Baud Rate 115200 Echo Disabled
Data Bits 8 Timeout(s) 10

Figure 2.8 - STM32 Flash Loader Configuration

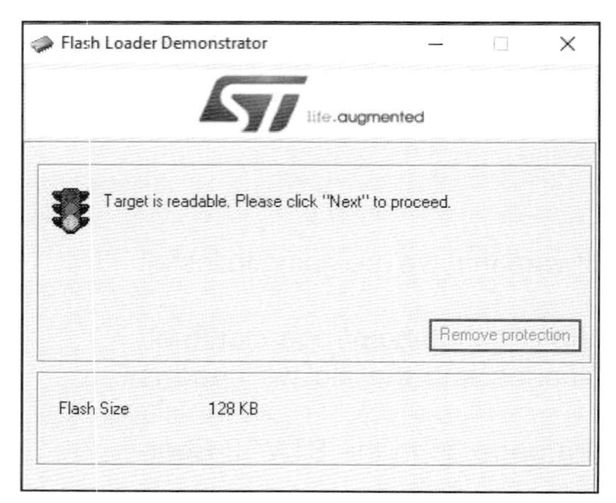

Figure 2.9 - STM32 Target Accessible

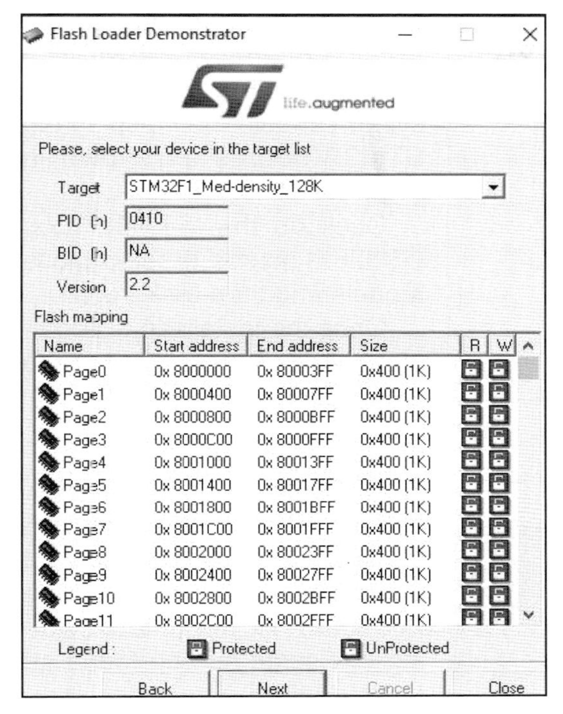

Figure 2.10 - STM32 Target Details

reset or removing power. Failure to complete this final step will not retain the bootloader code on the board and you will have to repeat the process above.

Step 6 – Installing the STM32 Arduino core

To enable support for the STM32 boards we need to install the core files into the Arduino IDE. To do this we need to start the IDE and open preferences and add this URL:

http://dan.drown.org/stm32duino/package_STM32duino_index.json as an additional board manager URL as shown in **Figure 2.13**.

Click OK to close the preferences window and then open the Arduino Board Manager by using the Tools menu, then select Board and then Boards Manager. This selection is shown in **Figure 2.14**.

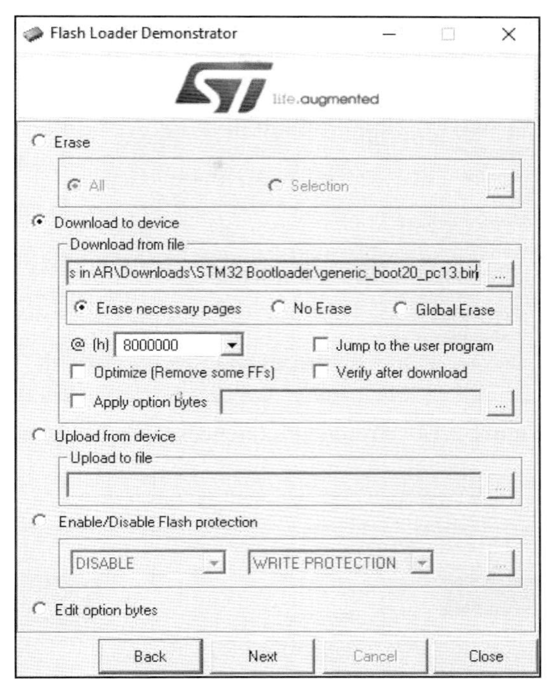

Figure 2.11 - Binary Bootloader File Selection

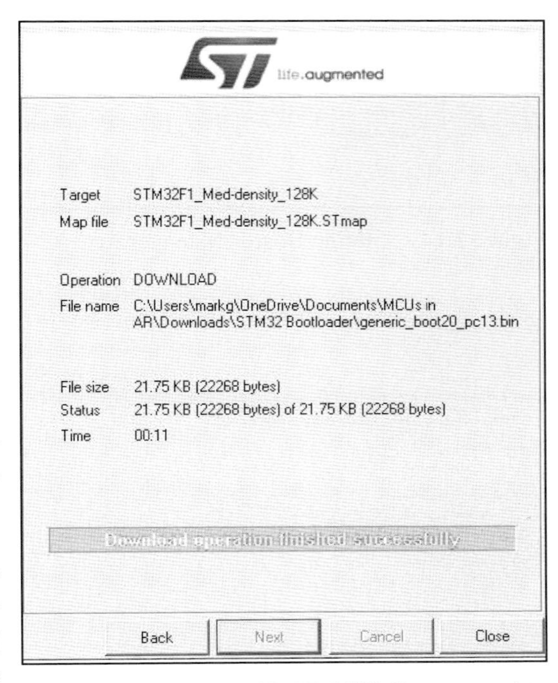

Figure 2.12 - STM32 FLASH Programming Complete

Figure 2.13 - Arduino Preferences Update

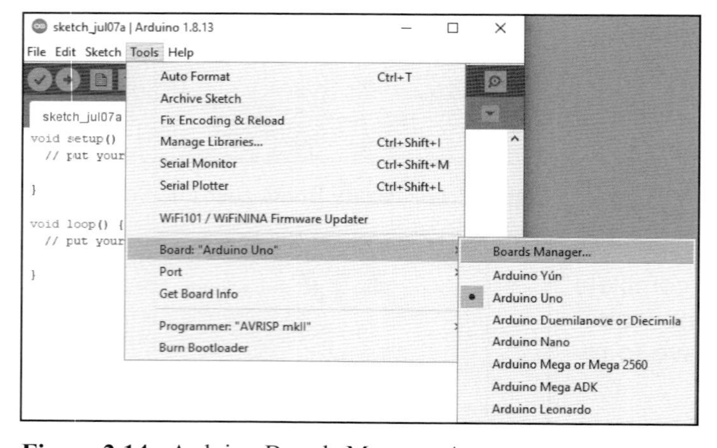

Figure 2.14 - Arduino Boards Manager Access

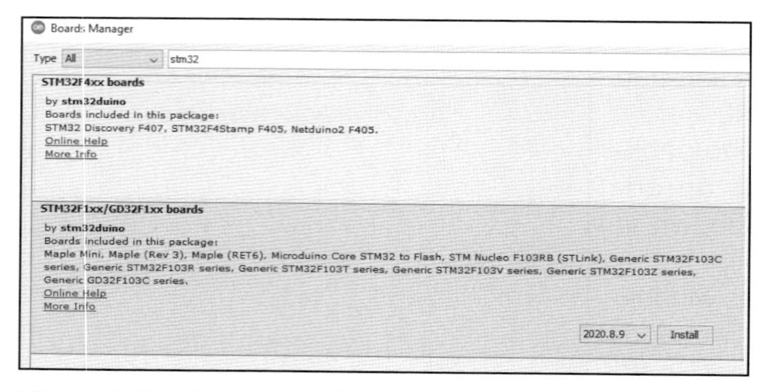

Figure 2.15 – STM32 boards

Once you have opened the Arduino Boards Manager, type stm32 into the search box. The system should then display the STM32 boards as shown in **Figure 2.15**.

Click on the 'STM32F1xx/GD32F11 boards' to select those models and then click Install.

Once the installation is complete click on close to return to the Arduino IDE.

Step 7 – Driver Configuration

It's now time to run a script on the PC to associate the correct driver files with the STM32 Blue Pill board. At the time of writing this script can be found at: https://github.com/rogerclark-melbourne/Arduino_STM32/tree/master/drivers

You should open the win directory and then download and run the "install_drivers.bat" file contained within. This is NOT the install_STM_COM_drivers.bat file rather the install_driver.bat file for the "Maple Serial" hardware. If using Edge browser on Windows 10 – you will need to "Keep anyway" as Edge believes this to be a dangerous file. It is normal practice to right click on the .bat file once downloaded to your PC and select "run as administrator". Note that this script takes a little while to run.

The files are also contained in this books extra page url in Downloads\Maple Driver.

Step 8 – Hello World

The equivalent of the Hello World program for embedded software developers is called “Blinky”. This is usually the first program run on a new environment to check all is working as expected.

We can now disconnect our FTDI USB to Serial adapter board from the PC and from the STM32 blue pill board. For the first time we now connect the blue pill USB port directly to the PC. Once connected the PC should install a new device called “Maple Serial” and after the device is installed there should be a new entry within Ports section of Device Manager as shown in **Figure 2.16**.

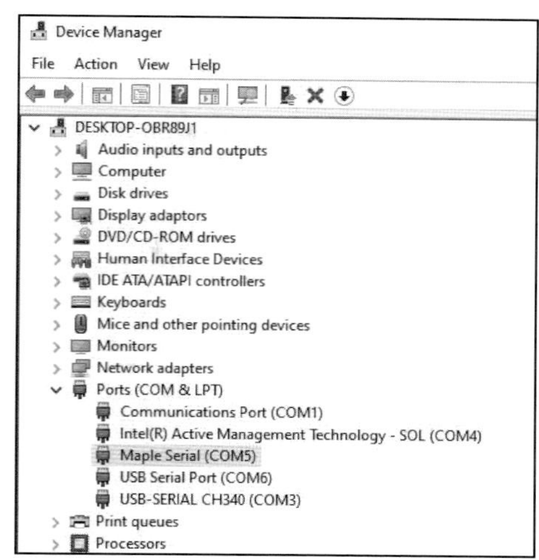

Figure 2.16 – Maple Serial entry in Device Manager

In the Arduino IDE, open the example sketch “Blink” from File -> Examples -> Basic -> Blink.

Configure the listed entries in the tools menu as follows:

- Board: Generic STM32F103 series
- Variant: STM32F103CB (20k RAM. 128k Flash)
- Upload method: STM32duino bootloader
- Port: Your Maple Serial COM port number

The Tools menu configuration a this point in the process is shown in **Figure 2.17**.

Once configured, you can command the IDE to upload the program. This can be done in several ways including using the shortcut CTRL-U. After the source code has compiled you should see the program upload to the board and the on-board LED should be flashing on and off at 1 second intervals.

Congratulations! You have just compiled and run your first embedded program on a STM32 target.

Let’s Talk About Blinky

In an embedded program, there are, generally speaking, two sets of things to be done. There are things that need to be done once at the beginning, like configuration and setup tasks, then there is the main job of the program itself that need to be done potentially forever.

All embedded programs will contain these two blocks of code. In the world of Arduino these are called “Setup” for the code that runs once at startup and “loop” for the code that runs continuously.

Board: "Generic STM32F103C series"
Variant: "STM32F103CB (20k RAM. 128k Flash)"
Upload method: "STM32duino bootloader"
CPU Speed(MHz): "72Mhz (Normal)"
Optimize: "Smallest (default)"
Port: "COM5"
Get Board Info

Figure 2.17 - Tools Menu Settings

If we look at our Blink example sketch we can see that this code is contained within the startup routine:

```
pinMode(LED_BUILTIN, OUTPUT);
```

This is the only configuration needed for our simple LED flashing program. This tells the system that we are using the built in LED pin as an output. The Arduino IDE already knows a whole bunch of things about our board, including the fact that the on-board LED is connected to pin PC13. PC13 means its General-Purpose IO (GPIO) port C (there are three GPIO ports on our MCU, ports A, B and C) and on Port C the LED is connected to pin 13. All other configuration needed to get the MCU up and running like the clock configurations are handled for us by the Arduino IDE. This is great when we know exactly how our MCU is connected to its PCB and on-board electronics as it saves the programmer a whole load of configuration work.

If we now move on to look at the "loop" part of our program:

```
digitalWrite(LED_BUILTIN, HIGH);
delay(1000);
digitalWrite(LED_BUILTIN, LOW);
delay(1000);
```

In these lines of code, we are telling the MCU to set the built in LED pin (as above we know that this is actually pin PC13) to HIGH which in our case is 3V3.

The delay command does what it says and simply waits for the duration in milliseconds placed within the brackets. In this case 1000 ms is one second.

Then we set the LED pin low which is 0V and finally we delay one second again.

Because these statements are contained in the loop function which is the code that runs continuously, these statements just repeat over and over. You could also attach a multimeter or oscilloscope to pin PC13 on your breadboard and see what's going on. Why not give it a try?

You could change the frequency that the LED flashes, I'll let you figure out how to do that.

Summary

Hopefully you will have successfully followed the steps above and now have a STM32 Blue Pill development board with an Arduino bootloader installed. You also have the Arduino IDE installed and configured on your PC to support STM32 boards and have run your first embedded program on a STM32 target.

We can now move on to Amateur Radio related topics and tackle our first project.

3

STM32 Project
A Simple Signal Generator

Shopping List

This section of the book uses the following components (Please refer to the hardware list file on the Books extra page for this book to obtain up-to-date URLs for where to buy these items):

- AD9833 Signal Generator Module
- Breadboard
- Hook-up Wires

Introduction

As our first project we are going to create a simple signal generator using the Analog Devices AD9833 chip. This device is a programmable waveform generator capable of Sine, Square and Triangular waveforms over a frequency range up to about 12.5 MHz.

It uses a simple three wire SPI (Serial Peripheral Interface) to connect to a controlling MCU (the Blue Pill board) which will be used to configure and control the waveform generator.

Project Configuration

Wire up your AD9833 and STM32 Blue Pill as shown in **Figure 3.1**.

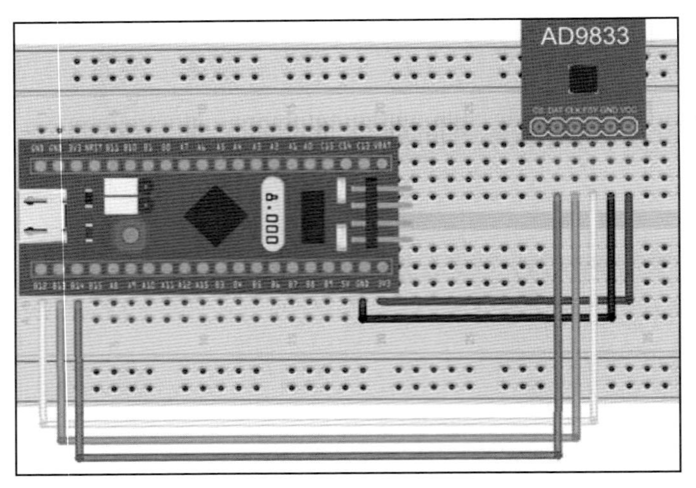

Figure 3.1 - Simple Signal Generator

Project Support Files

We will use a library for the first time in this project, so it is worth taking a moment to explain the concept in a little detail.

A library can be considered as a collection of someone else's code. In the world of Arduino there are libraries available for a vast array of external sensors and other devices all available in the public domain. In this project we want to interface our MCU with the AD9833 and could write appropriate software from scratch by referencing the device data-sheet, or we can simply let someone else do that for us and use their code!

From the Arduino IDE go to the Tools menu and select Manage Libraries. In the window that opens search for AD9833 and you should find a library called MD_AD9833 as shown in **Figure 3.2**, click on this library to select it and then click install.

The source code for this project is contained in the Books extra page under MCUs in AR\Arduino\AD9833_Basic_Test

Open the .ino file contained in the directory shown above using the Arduino IDE.

You will see that the first line of code in the source file is:

```
#include <MD_AD9833.h>
```

It is telling the Arduino IDE environment that we wish to use this library within our code.

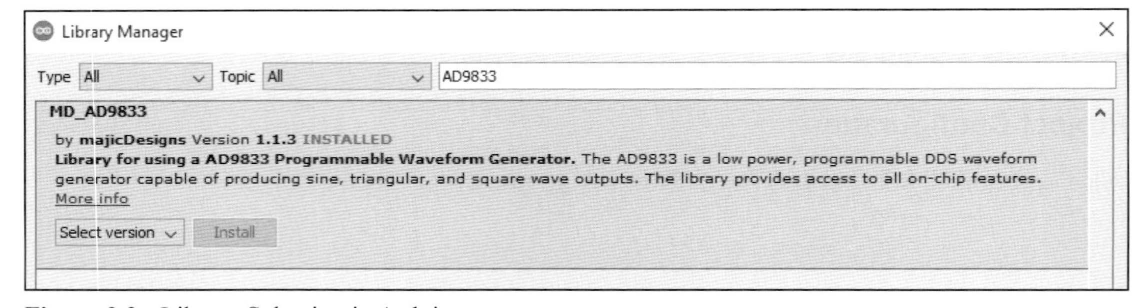

Figure 3.2 - Library Selection in Arduino

Very generally speaking, a library will consist of at least 2 files. There is always a .h file which is known as the header file. This .h file tells the outside world what the library contains that can be used externally – an interface definition if you like. The actual contents of the library and all the associated code will be contained in a second source file.

If you ever need to know what functions are offered by a library, just look in the .h file, everything you can utilise as a user of the library will be advertised there.

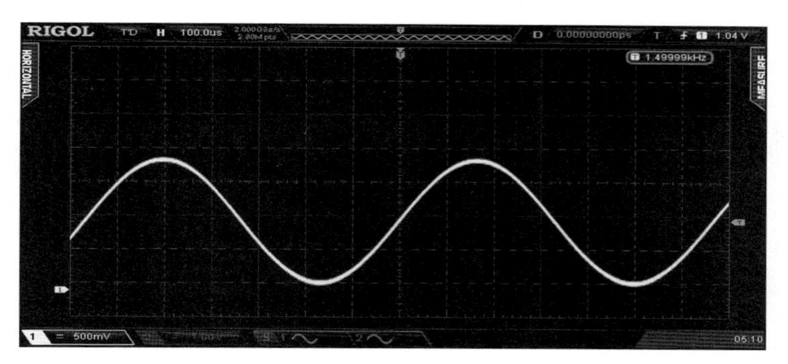

Figure 3.3 - Example Code Sine Wave Output

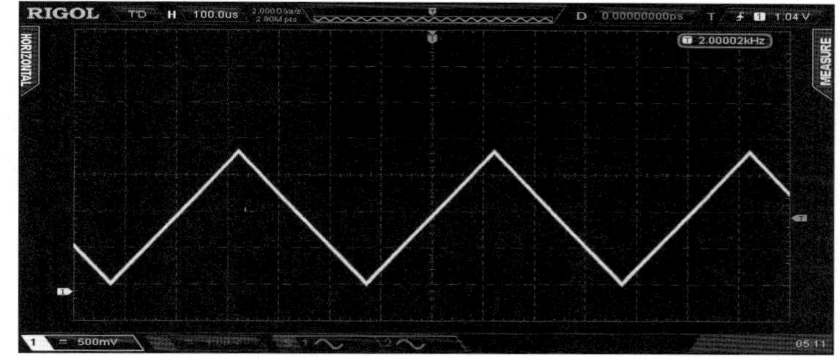

Figure 3.4 - Example Code Triangle Wave Output

Project Build

Once you have installed the MD_9833 library and opened the source code, we can now go ahead and build our project. Use the CTRL-U shortcut to compile and upload the project to the STM32 target chip and, once the upload has completed, you should have output from our AD9833 waveform generator. An oscilloscope would be ideal to monitor the waveform output, but a small loudspeaker will also let you hear the generator in action.

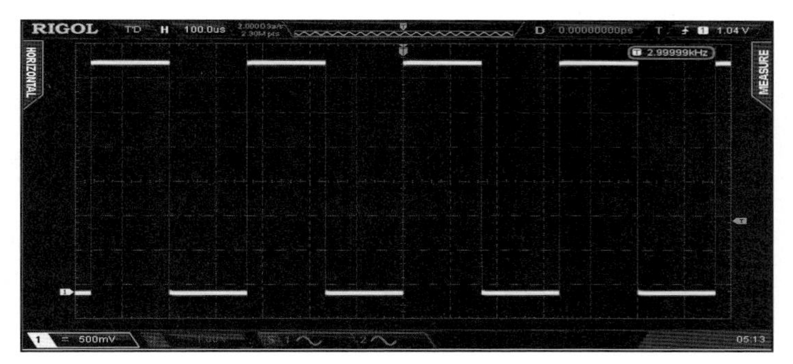

Figure 3.5 - Example Code Square Wave Output

The simple test code is designed to run through the following sequence repeatedly:

1. Sine wave at 1.5kHz for 3 seconds
2. Triangle wave at 2kHz for 3 seconds
3. Square wave at 3kHz for 3 seconds

A look at the code

Anything that follows a // within the code is a human readable comment and is ignored by the compiler. It is always a good idea to comment your code so that when you return to it later, you have ample reminders of the functionality and your intent while writing it.

Global Space

In the source file we have some lines of code at the top of the file below the library include command. These define global variables which, in our example project, are simply defining which MCU pins the AD9833 is connected to. Global variables are accessible by all other parts of our code.

```
// Pins for comms with the AD9833 IC
const int DATA        = PB14;  // Data
const int CLK         = PB13;  // Clock
const int FSYNC       = PB12;  // Load (FSYNC pin on
                               // AD9833)

MD_AD9833 SigGen(DATA, CLK, FSYNC); // use pins defined above
                                    // for interfacing
```

We are defining in software that:

- The AD9833 DAT pin (data) is connected to Blue Pill pin PB14
- The AD9833 CLK pin (clock) is connected to Blue Pill pin PB13
- The AD9833 FSY pin (FSync) is connected to Blue Pill pin PB12

Then we create an instance of the AD9833 library functions and call it SigGen and define which pins of the AD9833 are connected to which of our Blue Pill board's pins.

Initialisation

In our example project there is only one line of initialisation, contained in the setup routine:

```
SigGen.begin();
```

This line of code is a call to a procedure contained within our AD9833 library that we have created an instance of called SigGen and the procedure itself is called begin.

The begin procedure in the library is called to perform the necessary initialisation of the AD9833 chip. We don't need to worry about the detail of this initialisation; it's all taken care of for us by the library.

Main Loop

Our main loop contains the following lines of code:

```
// set the frequency on the main channel to be 1500Hz
SigGen.setFrequency(MD_AD9833::CHAN_0,1500);
// set the output type to be Sine
SigGen.setMode(MD_AD9833::MODE_SINE);
// hold the boat for 3 seconds
delay(3000);

// set the frequency on the main channel to be 2000Hz
SigGen.setFrequency(MD_AD9833::CHAN_0,2000);
// now set the output type to be a triangle
SigGen.setMode(MD_AD9833::MODE_TRIANGLE);
delay(3000);

// set the frequency on the main channel to be 3000Hz
SigGen.setFrequency(MD_AD9833::CHAN_0,3000);
// and now a square wave
SigGen.setMode(MD_AD9833::MODE_SQUARE1);
delay(3000);

The first line of our main loop:

SigGen.setFrequency(MD_AD9833::CHAN_0,1500);
```

Line 1 calls a procedure called setFrequency within our SigGen library routine and passes it two parameters. The first parameter is the channel of the waveform generator to use, the second is the frequency in hertz.

We tell the waveform generator what type of waveform we want (the options available as declared in the .h file are SINE, SQUARE1, SQUARE2, TRIANGLE):

```
SigGen.setMode(MD_AD9833::MODE_SINE);
```

In this case we are telling the waveform generator that we want to produce a sine wave.

Then we execute a delay statement that we came across before, in the basic Blink program.

You should be able to see from the source code that the program goes through a sequence of different waveforms and frequencies with a delay between each before repeating.

Things to Try

1. Use the SQUARE2 waveform type – what does that do and how is it different from the SQUARE1 waveform
2. Alter the durations and frequencies of the waveforms.

4

STM32 Project
Controlling a DDS Oscillator

Shopping List

This section of the book uses the following components (Please refer to the hardware list file on the Books extra page for this book to obtain up-to-date URLs for where to buy these items):

- AD9850 DDS Module

Introduction

In our first project, we used a library to get the necessary code to control our waveform generator. In project 2 we will look in more detail at directly controlling an external device – in this example we will use an Analog Devices AD9850.

The AD9850 is an immensely popular DDS (Direct Digital Synthesiser) available built into modules that include the necessary external components to make a complete MCU controllable system.

When looking to write software to control any external device, the datasheet is an excellent place to start. In the case of the AD9850 this can be found at:

https://www.analog.com/media/en/technical-documentation/data-sheets/AD9850.pdf

The datasheet is also located in the Datasheets folder of the RSGB book file repository.

Starting on page 9 of the datasheet is some detailed information about programming the device including the necessary timing information. We can use the information provided to create a set of routines to help us control the DDS.

Writing the Control Routines

To control all the DDS functionality, we need to create several routines that we can call from our setup (for initialisation) and main loop (for operation).

The key functional items for the DDS operation are:

- Initialisation
- Enable Output at given Frequency & Phase
- Turn Off

Studying the datasheet, it is clear we will also need to have routines to:

- Pulse digital lines high & low
- Transfer data to the DDS a byte at a time

Defining the DDS Connections

As was the case with the waveform generator, we will first need to define which pins of the DDS are to be connected to our STM32 Blue Pill board. The connections we need to define are RESET, DATA, FQ_UD (this is a load pin) and CLOCK.

We will use the following lines of code to define these connections:

```
const int DDSRST      = PB8;
const int DATA        = PB7;
const int FQ_UD       = PB9;
const int CLOCK       = PB6;
```

As we are not using anyone else's code, we need to tell the STM32 MCU that we wish to use these specific pins as digital outputs (which can be 1s or 0s) and we do this in the setup routine, as part of our initialisation process:

```
pinMode (DATA, OUTPUT);    // DDS pins as output
pinMode (CLOCK, OUTPUT);
pinMode (FQ_UD, OUTPUT);
pinMode (DDSRST, OUTPUT);
digitalWrite(DATA, LOW);   // internal pull-down
digitalWrite(CLOCK, LOW);
digitalWrite(FQ_UD, LOW);
digitalWrite(DDSRST, LOW);
```

Each pinMode statement tells the MCU that the pin is to be used as a digital output and the digitalWrite statement configures the GPIO pin to use an internal pull down defaulting the output to LOW.

Pulsing Lines High and Low

To pulse a digital line high then low, as required by the datasheet, we can create a routine in our software:

```
void pulseHigh(int pin)
{
  digitalWrite(pin, HIGH);
  delayMicroseconds(2);
  digitalWrite(pin, LOW);
}
```

This defines a procedure we can call from elsewhere where we will pass as a parameter the pin number we wish to toggle, the code will then set the pin HIGH (3V3) wait 2 micro-seconds and then set the pin LOW (0V).

Transferring Data to the DDS

TransfterByte

The datasheet explains that we need to send each byte of data to the DDS one bit at a time pulsing the CLOCK line after each bit. We can write a routine to perform this task for us:

```
void TransferByte(byte data)
{
  for (int i=0; i<8; i++, data>>=1)
  {
    digitalWrite(DATA, data & 0x01);
    pulseHigh(CLOCK);
  }
}
```

There is some fiendishly clever jiggery-pokery going on here, so it's worth spending a little time explaining this routine in more detail.

The routine is called TransferByte and we can call this from elsewhere in our software and pass it some data. The data we pass is of the standard type byte which means it is 8 bits in length and can have a value in the range 0 to 255 decimal (which is 0 to FF in hex).

The routine then executes a loop using the C language for statement. The for statement will make the code contained within the following open and closed brackets execute 8 times, each time the loop repeats the value stored in the data that was passed to the routine is shifted to the right by 1 bit.

For example, if we were to call the TransferByte routine from our code and pass it data containing the value of decimal 9, this would be 09 in hex and:

00001001

as binary. The first time round our loop the value in data would remain as binary 00001001 but the second time round the loop it would be shifted to the right by one bit and would become:

00000100

In our example we need to send data one bit at a time, and this for statement will place the bit we need to send, each time we loop round, into the right most bit of the data value.

We then use the statement:

```
digitalWrite(DATA, data & 0x01);
```

to set the value of the digital output pin DATA (defined earlier as PB7) to the value in the right most bit of the data variable.

The is achieved by this part of the statement:

```
data & 0x01
```

which is using the C language and function (&) to compare the data with decimal 1 (hex 01 or binary 00000001). The result of this and function will be the value in the right most bit of our data. Depending on the result of the and function, the digital output pin DATA will be set to either HIGH (3V3) or LOW (0V) replicating the value of the right most bit in our data of 1 or 0 respectively.

After each data write we pulse the CLOCK line as specified in the datasheet.

We repeat this trick 8 times, once for each of the 8 bits in our byte of data we want to send to the DDS.

Supporting Routines

We now create several supporting pieces of code that utilise the routines we have defined above:

UpdateDDS

```
void UpdateDDS(int32_t freq, uint8_t Phase)
{
  for (int b=0; b<4; b++, freq>>=8)
  {
    TransferByte(freq & 0xFF);
  }
  TransferByte (Phase & 0xFF);
  pulseHigh(FQ_UD);
}
```

UpdateDDS is a routine that can be called from our software and passed a frequency and phase value. This will then perform the necessary data transfer to the DDS for that target output frequency and phase. The routine has a loop which is executed 4 times and the data being sent uses a similar bit shifting technique to TransferByte above, but this time it is shifted by 8 bits (one byte) in each loop iteration. The routine passes the data one byte at a time to the TransferByte routine and also pulses the FQ_UD (load) pin of the DDS.

SetFrequency

```
void SetFrequency (uint32_t TargetFrequency, uint8_t
TargetPhase)
{
 int32_t freq = TargetFrequency * 4294967295/DDSClock;
  uint8_t phase = TargetPhase << 3;
  UpdateDDS (freq, phase);
}
```

The SetFrequency routine utilises the UpdateDDS code defined above and takes a frequency and phase in Hz and performs the mathematics on the real-world values defined by the datasheet before calling UpdateDDS.

DDSClock is defined in the software:

```
const uint32_t DDSClock = 125000000;
```

and is the frequency in Hz of the on-board clock used to drive the DDS. In my module this is 125MHz. If you find your DDS frequencies slightly off frequency, you could tweak this value as a means of calibration.

The value sent to the DDS to set the frequency is:

frequency in Hz * 232 / Clock Frequency and is accomplished by the line:

```
int32_t freq = TargetFrequency * 4294967296/DDSClock;
```

and the value representing the phase of the signal is the raw value shifted left by 3 bits in accordance with the datasheet.

```
uint8_t phase = TargetPhase << 3;
```

The resultant values in variables freq and phase are then passed to the DDS using the routine DDSUpdate.

DDSDown

```
void DDSDown()
{
  pulseHigh(FQ_UD);
  TransferByte(0x04);
  pulseHigh(FQ_UD);
}
```

The DDSDown routine simply follows the instructions in the datasheet to stop the DDS from operating by passing the value 04 hex (also 4 in decimal) to the DDS with the load line (FQ_UD) pulsed before and after the data transfer.

Summary

We can now use the routines we have created above to make a working example of DDS programming. The source code for this example is in the Books extra page in Arduino\AD9850 Routines\DDS9850_Routines_STM directory and is called DDS9850_Routines_STM.

In our example we will program the DDS to output a frequency of 7.028 MHz. Following the maths specified in the datasheet:

Data to Send to DDS = (Target Frequency * 232) / DDS Clock Frequency

Therefore, for our example:

Data to Send to DDS = (7028000 * 4294967295) / 125000000 = 241480241 or 0E 64 B2 31 Hex

In accordance with the datasheet the value 0E 64 B2 31 Hex must be sent to the DDS as separate bytes in reverse order followed by a pulse of the FQ_UD (load) line.

Project Configuration

Wire up your AD9850 module and STM32 Blue Pill as shown in **Figure 4.1**.

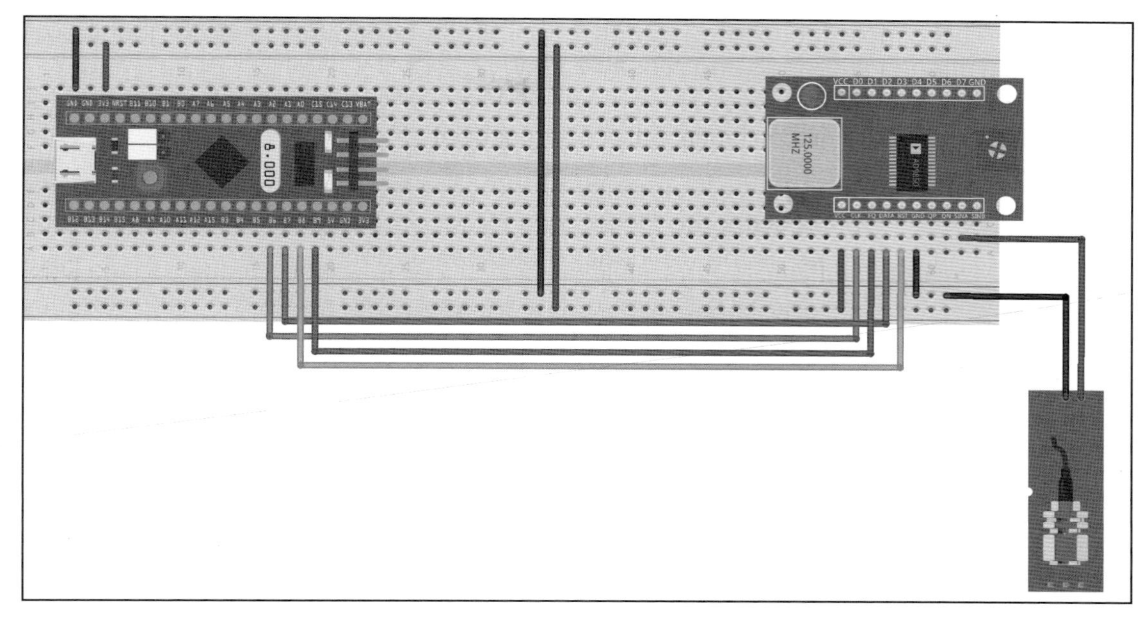

Figure 4.1 - AD9850 Configuration

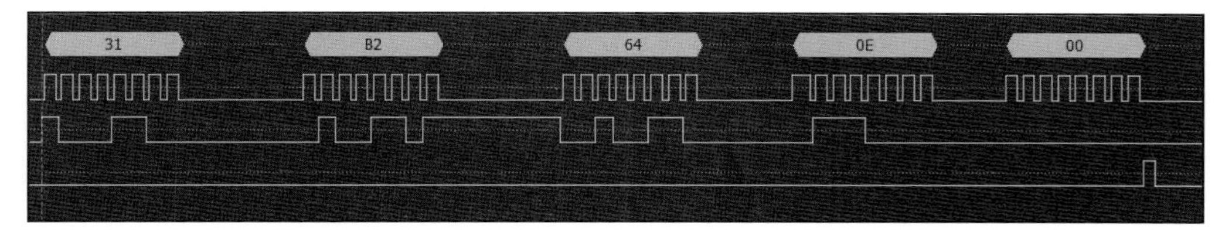

Figure 4.2 - Logic Analyser Capture of Data to DDS

Project Build

Once you have opened the source code we can now go ahead and build our project. Use the CTRL-U shortcut to compile and upload the project to the STM32 target and, once upload is complete, you should see an output from the DDS at 7.028 MHz.

An oscilloscope or frequency counter would be ideal to monitor the output, but you could find the signal on a nearby HF receiver if you have one.

Looking at the actual data sent by the software to the DDS in the authors example, you can see the expected values decoded by a logic analyser in **Figure 4.2**, the fifth byte of data being sent is the phase value of 0 degrees. The top line is the CLOCK, the second is the DATA and the third FQ_UD (load).

This simple test code generates a single output at this fixed frequency but is written in a generic and reusable way.

Things to Try

1. Change the frequency of the generated signal – what frequency range does the DDS cover? Can you see any differences in the signal as the frequency increases or decreases?
2. Use the DDSDown routine in combination with the setFrequency and delay functions to turn the DDS on and off.

5

STM32 Project
A DDS based VFO for the Super Sudden

Shopping List

This section of the book uses the following components (Please refer to the hardware list file on the Books extra page for this book to obtain up-to-date URLs for where to buy these items):

- Keyes rotary encoder
- ILI9341 TFT display

Introduction

Our first projects introduced us to some basic concepts in code writing and controlling external devices. In this project we will put these concepts together with some other devices to make a working VFO which we will use with a well-known receiver design.

We will use a rotary encoder and a TFT display to create a tuneable VFO and make a HF receiver.

A rotary encoder is a mechanical device which provides two digital outputs indicating its direction & speed of rotation. One of the outputs lags behind the other, and therefore the phase difference in the outputs determines the rotation direction and the frequency of the output signal determines the rotation speed.

In an encoder there are normally two contact pins A and B plus a set of evenly spaced contacts which rotate and are connected to the common pin C. Imagine the common pin is at our VCC voltage of 3V3 and the output pins are tied to ground as illustrated in **Figure 5.1**.

As the disk starts to rotate clockwise, output pin A will contact the common VCC connection before pin B changing their states from low to high, generat-

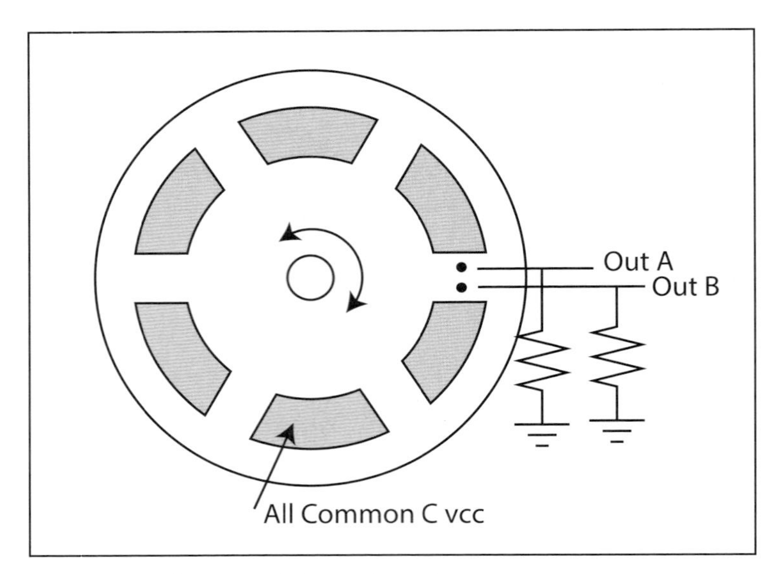

Figure 5.1 - Conceptual Rotary Encoder

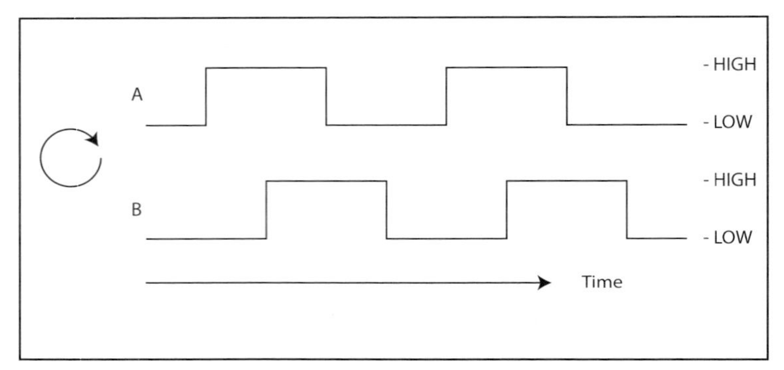

Figure 5.2 - Conceptual clockwise waveform

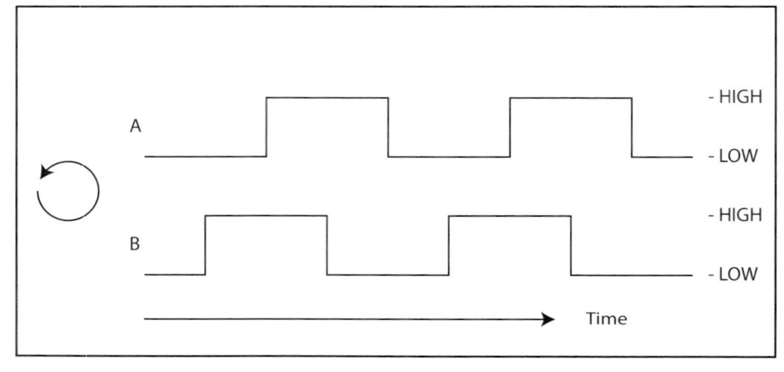

Figure 5.3 - Conceptual anti-clockwise waveform

ing signals like those shown in **Figure 5.2**.

Likewise, when turned anti-clockwise output pin B will contact the common VCC connection before pin A, generating signals like those shown in **Figure 5.3**.

Therefore, our MCU can monitor the signals from the rotary encoder and determine rotation direction and if required calculate rotation speed.

Rotary encoders can be electrically very noisy, and care needs to be taken to debounce the signals properly and ensure that only genuine rotations are acted upon. The code in our example takes care to cater for this real-world situation by having a reference of valid previous and current states of the output pins.

Project Configuration

Wire up your AD9850, Blue Pill, Rotary Encoder and TFT display as shown in **Figure 5.4**.

Project Support Files

We are going to use a library to connect to our TFT display, but this library has been modified to work with the STM32 core. Instead of installing the library (as we did with the AD9833) we will use a local copy.

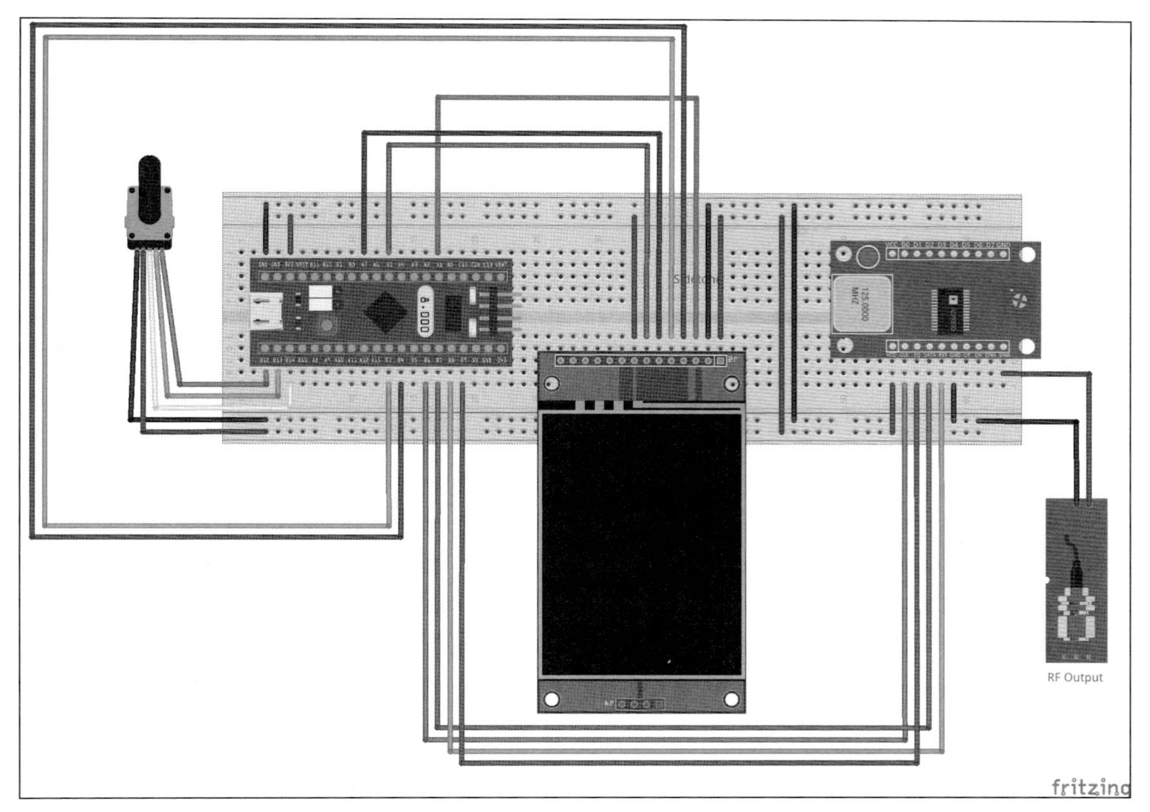

Figure 5.4 - DDS VFO

All the source code including the local copy of the ILI9341 is contained in the Books extra page in the directory Arduino\AD9850_vfo_STM32_ILI9341. You need to copy the entire directory to your Arduino source folder.

You will see when you open the .ino source file, that the #include statements use a slightly different syntax to those we used before. This syntax tells the compiler that the files are in the source code directory rather than the library store.

Project Build

Once you have copied the source repository to your local Arduino source folder, you are ready to build the project. Open the .ino file in the Arduino IDE and use the CTRL-U shortcut to compile and upload the project to your STM32 board. Once completed you should see the display become active and have an output from the DDS.

Turning the encoder will tune the VFO between the limits defined in the code. This project has been used by the author as the VFO for the "Supper Sudden" HF receiver.

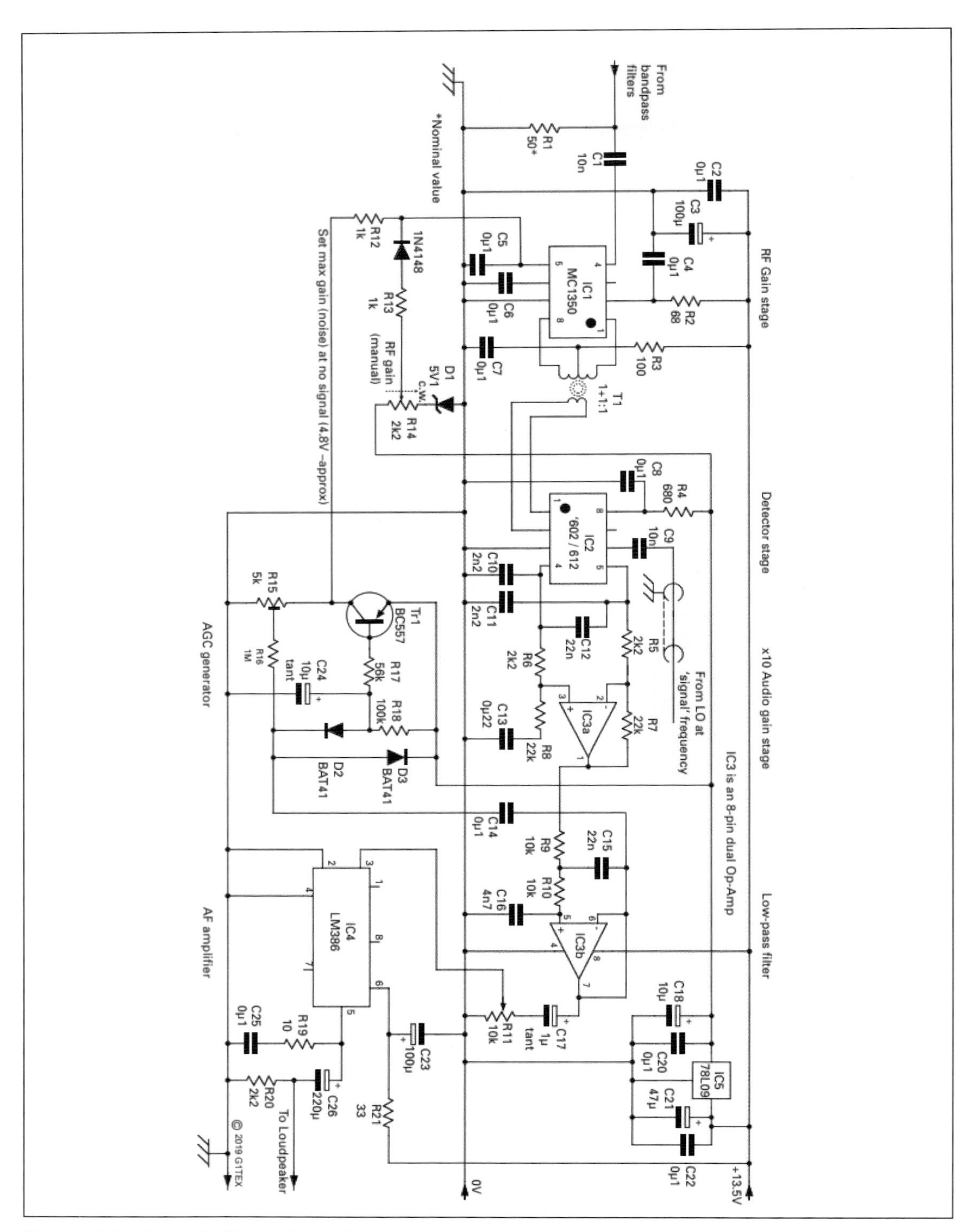

Figure 5.5 - Super Sudden Schematic

The original "Sudden" receiver was first published by the late, great, George Dobbs G3RJV. The first publication of a Sudden receiver was in March 1991 and various modifications and improvements have been published since. The most recent modification I am aware of was in the GQRP journal "Sprat", issue 181, which contained a "Super Sudden" by Charles, G1TEX. This version is available as a kit from Spectrum Communications which requires the addition of a band pass filter and VFO to make a complete receiver.

Designs for suitable band pass filters are available on the website of the G-QRP club in the public technical pages and my build uses their 80M variant.

The schematic below of the receiver in **Figure 5.5** is courtesy of G1TEX.

The authors build of the Super Sudden with STM32 DDS is shown **Figure 5.6**.

Figure 5.6 – Super Sudden Receiver

A look at the code

Our DDS routines from the previous project are copied into this source file without any changes. They are called from the other routines in the software as needed to control the DDS. Note that the DDSDown routine has been excluded as it is not required in this project.

We have some new routines in the source code to handle the rotary encoder, we also use the new library functions for controlling the TFT display. The TFT routines are straightforward to use and the library makes several different facilities available for our use. For example:

```
tft.setRotation(1);
tft.fillScreen(ILI9341_BLACK);
tft.fillRect(0, 110, 320, 95, ILI9341_RED);
tft.setCursor(110, 10);
tft.setTextColor(ILI9341_RED);
tft.setTextSize(4);
tft.println("G-QRP");
tft.setTextColor(ILI9341_GREEN);
```

```
tft.setCursor(55, 60);
tft.setTextSize(3);
tft.println("Super Sudden");
```

The code above performs the following using routines from the TFT library:

- Set the screen to horizontal mode (it can be used anyway up you like)
- Fill the entire screen with Black
- Create a rectangle starting at coordinates 0 (x – across), 110) (y – down) and 320 pixels wide and 95 pixels high and paint it Red.
- Set the text cursor to be at coordinates 110 (x-across), 10 (y-down)
- Set the text colour to be Red
- Set the text size
- Display the text G-QRP
- Set the text colour to be Green
- Set the text cursor to be at coordinates 55 (x-across), 60 (y-down)
- Set the text size
- Display the text Super Sudden

Things to Try

1. Change the default frequency along with the VFO upper and lower limits to work on the 40M amateur band.
2. Change the Display colours and layout to your taste – also remove G-QRP from the graphics and replace with your callsign if you have one.
3. Study the rotary encoder routines and understand how they work.
4. Place the rotary encoder routines into your own library files for use in your own future projects.

6

STM32 Project
A Simple DDS based CW TX

Introduction

The QRZ.com pages of W6JL contain some excellent information on projects using a combination of the old and the new. Don uses older technology for QRO power, combined with more modern DDS methods for CW generation.

I was interested in his CW keying circuit which he claims, “is nothing new”, but none the less, I have used it as the inspiration for part of the project I present here.

Ideally, we want to achieve about 5 ms or so rise and fall times for good quality keying. Anything less tends to make audible keyclicks away from the carrier frequency, although that may be no big deal if not running QRO, wherever we can, we should strive to construct good quality homebrew equipment and keep our bands clean.

The work of W8JI explains the concept of key-clicks well:

https://www.w8ji.com/what_causes_clicks.htm

In summary, what we are going to do here, is create a simple DDS based CW transmitter, with a key shaping circuit, that can drive a linear amplifier.

We will use our DDS and associated routines, the TFT, our STM32 board plus some external electronics to create the project.

Project Configuration

Wire up your AD9850, STM32 Blue Pill board & TFT screen in accordance with **Figure 6.1**.

We also need to construct some external electronics to provide the CW key shaping and a class A amplifier with LPF to boost the DDS output. The schematic

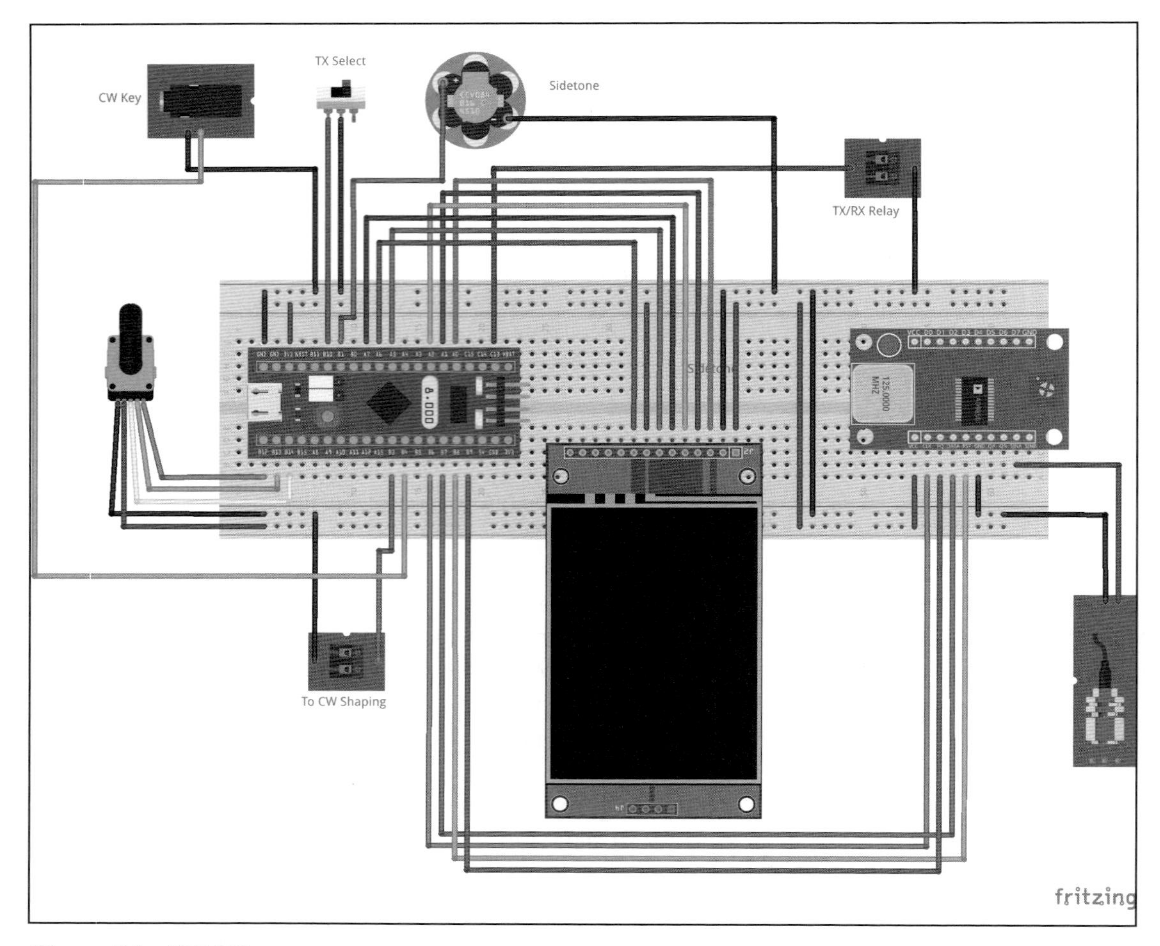

Figure 6.1 - CW TX

for this is presented in **Figure 6.2**. Note that the LPF in the schematic is for 80M.

In my build of this project I used an SDR as the receiver to accompany the TX, so a TX/RX relay is included in the design, switched by the software, to route the antenna to the receiver during RX and ground the RX antenna during TX.

A word about PWM

Pulse Width Modulation (PWM) is a mechanism used by MCUs to reduce the average power delivered to a connected peripheral by chopping it into discrete components. This can be used for speed control of motors and is a common mechanism used in robotics. Instead of having a signal that is either HIGH (3V3) or LOW (0V) as we have in past projects, here we use PWM for the first

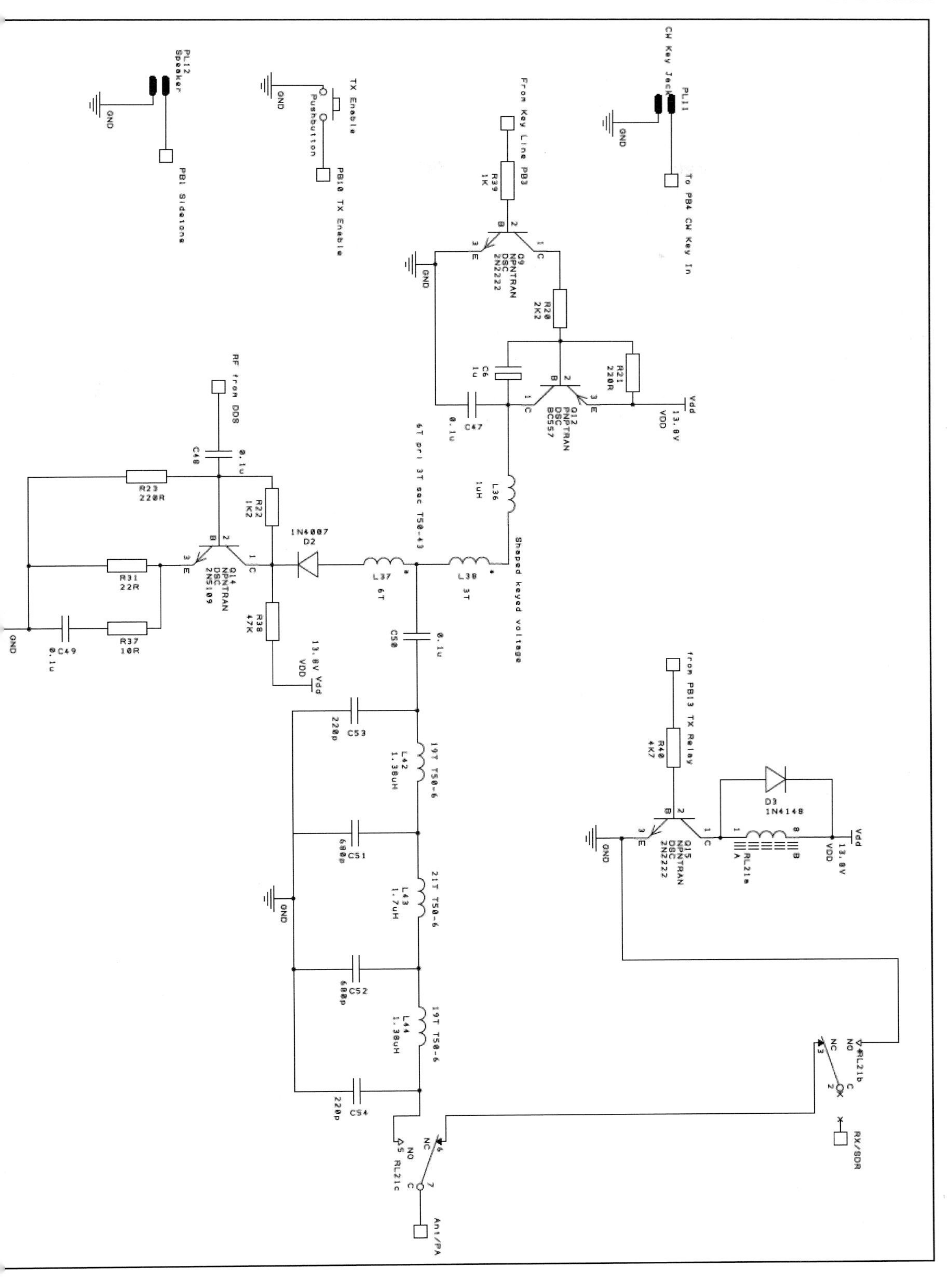

Figure 6.2 - CW TX Amp and Key Shaping

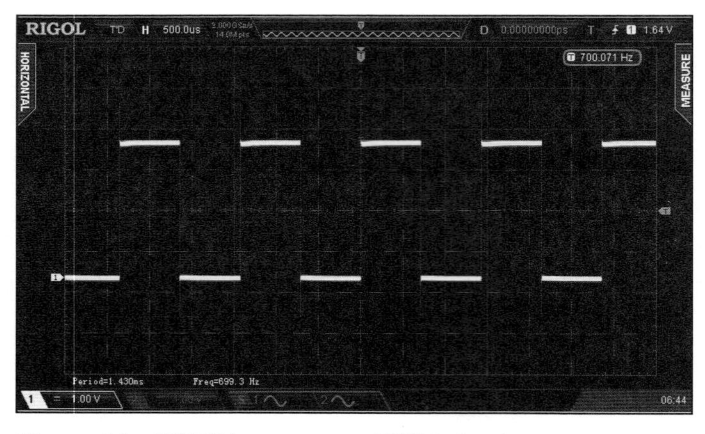

Figure 6.3 - CW Sidetone output PWM signal

time. In this project, we will use the PWM feature of our MCU to generate audio.

The Arduino IDE provides us with some very convenient functions that we can use to generate PWM signals, including audio tones. In our project code we have the following statements:

```
const int SideTone = PB1;
const int CWPitch = 700;

tone(SideTone, CWPitch);
```

This will generate a 50% duty cycle output on pin PB1 with a frequency of 700 Hz; when coupled to a small loudspeaker or sounder it will produce a very acceptable CW sidetone. The code:

```
noTone(SideTone);
```

will switch the sidetone signal off. The output generated by the code above is shown in **Figure 6.3**.

Project Support Files

All of the source code and supporting files needed for the project are contained in the Books extra page in the directory: Arduino\AD9850_CWTX_STM32.

Project Build

Once you have copied the source repository to your local Arduino source folder, you are ready to build the project. Open the .ino file in the Arduino IDE and use the CTRL-U shortcut to compile and upload the project to our STM32 board. Once completed you should see the display become active and there will be an output from the DDS at 3.56MHz when the key is down.

The DDS is running permanently at the selected frequency all the time. When the software is in TX the RF is switched by the external electronics and key shaping circuit.

Things to Try

1. Change the CW Sidetone pitch to your preferred audio frequency
2. Change the software so the TX is on the 40M band

7

STM32 Project
A Shack Power / dBm Meter

Introduction

Accurately measuring power in dBm is a frequent requirement in my experiments with RF. Commercial power meters are expensive and presented here is a very useful and accurate homebrew alternative.

The heart of the project is a well-known device from Analog Devices; the AD8307. The AD8307 is an extremely useful SMD device that takes an RF input and converts it to a DC voltage logarithmically.

We can use these devices with a directional coupler to create a measurement of forward and reverse power, some mathematics will then generate the SWR reading. We can use this kind of setup between a transmitter and an antenna.

We can also use this device as a standalone 50-ohm instrument for accurately measuring an RF signal's amplitude allowing us, for example, to measure losses and gains in circuity.

We will provide both a directional power meter and a 50-ohm dBm meter in this project but tackle the project in stages with the directional power meter first.

The components for this home-brew design are available from several suppliers, one UK local supplier is Farnell UK (uk.farnell.com).

Sampling RF

To create the power meter part of our project we will need to homebrew a directional coupler. The schematic of my variant is shown in **Figure 7.36**.

Ideally this would be built in a screened enclosure with each of the four ports isolated from the others. My, very poorly constructed, implementation of this schematic is shown in **Figure 7.37**.

The RF sample from the directional coupler is then taken to the AD8307. The RF sensing circuit I have used (one for forward and one for reflected) is shown in **Figure 7.38**.

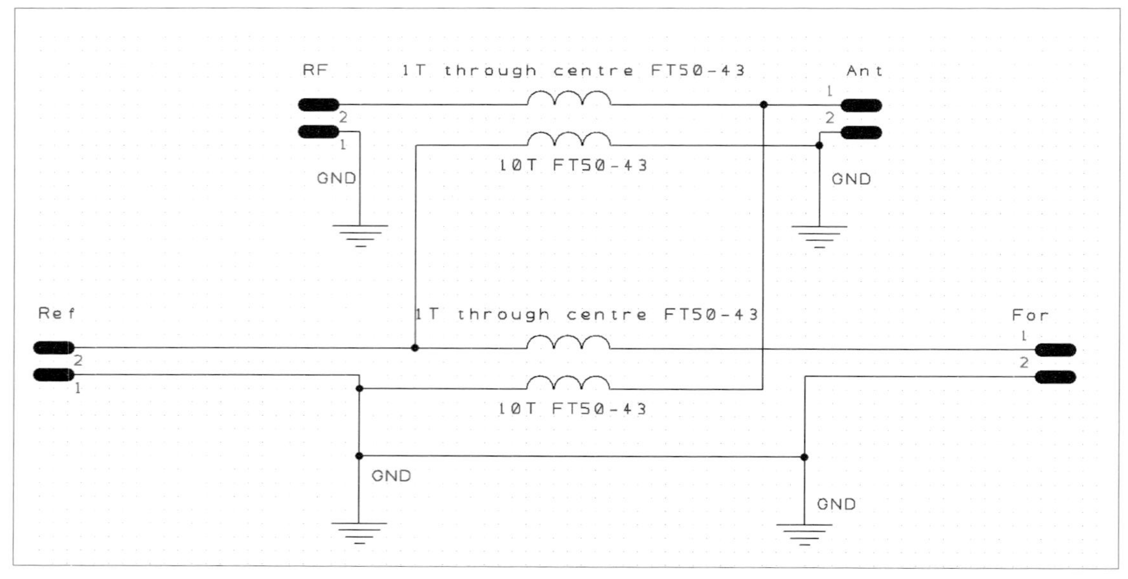

Figure 7.36 - Directional Coupler Schematic

Figure 7.37 - Directional Coupler Build

The output of the RF sense circuit is then fed to the analogue input of our STM32 board.

Using the RF Sample

After we have our RF sample fed to our Analogue to Digital converter (ADC) within our Micro Controller Unit (MCU), we need a mechanism to convert this to an accurate dBm reading.

I used my shack signal generator to create a series of RF samples that I fed through the RF sampling circuit. A simple piece of software was created to take 100 separate samples from the ADC (10 ms between each sample) to create a dataset, then we take the 85th percentile of that dataset and round it to the nearest 10. For each different amplitude and frequency, we received the resultant value which was then plotted using excel to create the graph shown in **Figure 7.39** .

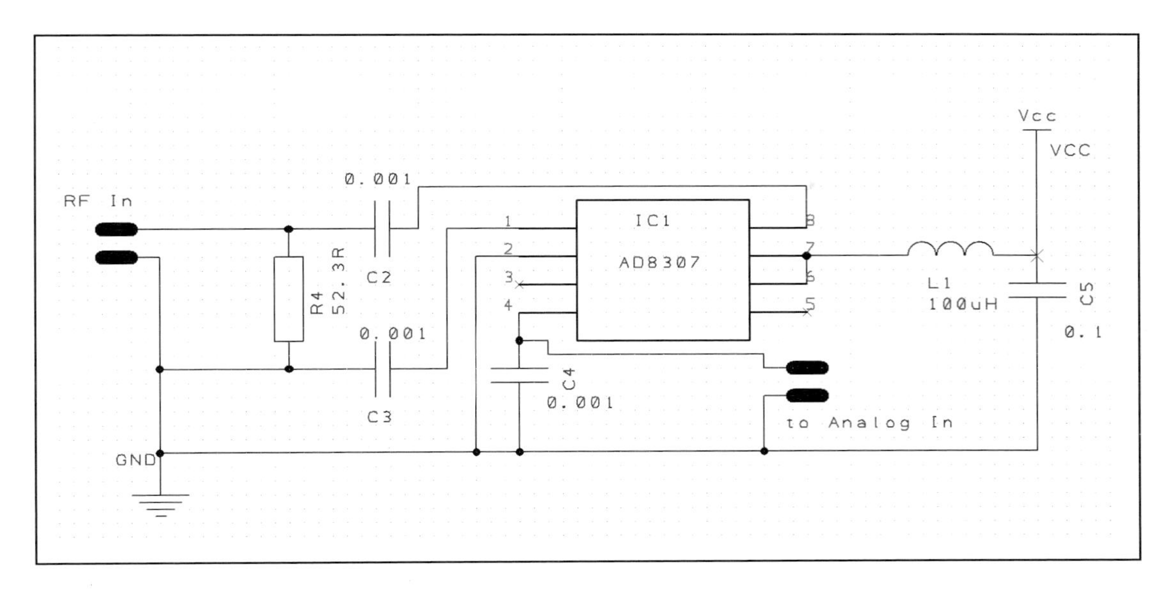

Figure 7.38 - RF Sense Circuit

Within Excel you can add a linear trendline to a graph plot and ask the software to show the equation on the chart.

Using the equation shown on the charts for the forward and reflected datasets, we can then calculate the dBm value for ANY ADC value.

For example, using the forward dataset (blue in the chart above), given an ADC value of 2500, the dBm value would be:

dBm = (0.0316 * 2500) – 87.185

Finally, to generate a usable dBm value I added an additional step to my algorithm, shown in **Figure 7.40** which now rounds to the nearest 0.5dB.

If we want to sample RF outside the range of the AD8307, we simply add an attenuator to the input. In my case I have added a 39dB attenuator to the forward RF sample circuitry and 21dB to the reflected.

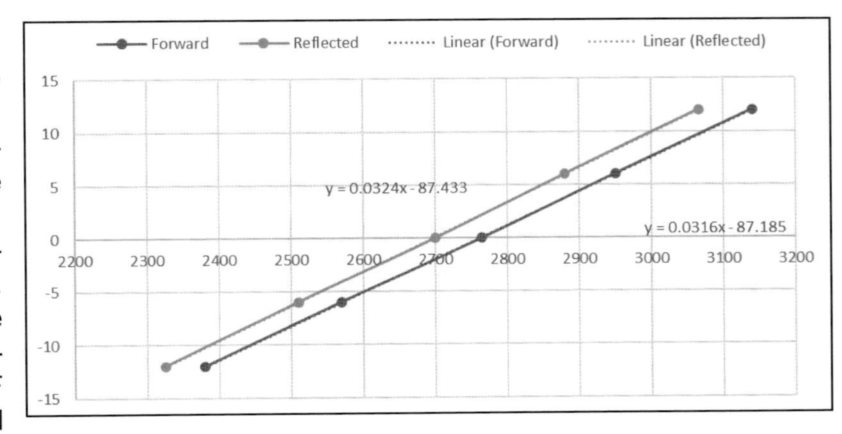

Figure 7.39 - Excel Graph of ADC Values

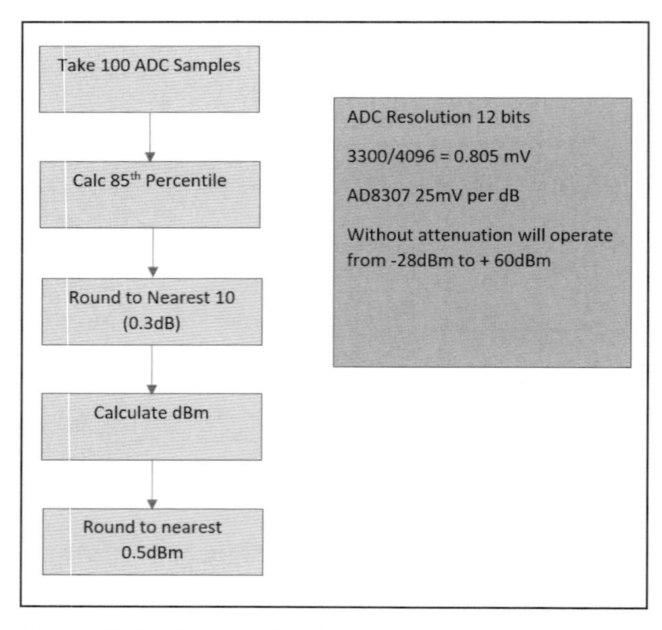

Figure 7.40 - Sample Algorithm

We can now hook up the Directional Coupler, the RF Sample circuitry and our MCU with a suitable display to create a forward and reflected reading power meter, we can even add the calculation for SWR using:

Where Pr = Reflected Power and Pf = Forward Power

The software takes the samples of forward and reflected power ever 10ms, then every 2 seconds it updates the display with the forward, reflected and SWR readings.

Project Configuration

Wire up your Directional Coupler, RF Sense circuits, Blue Pill and TFT by plugging them or their wires into the patch board in accordance with **Figure 7.41** .

Project Support Files

All of the source code and supporting files needed for this part of the project are contained in the Books extra page in the Directory: Arduino\PowerMeter

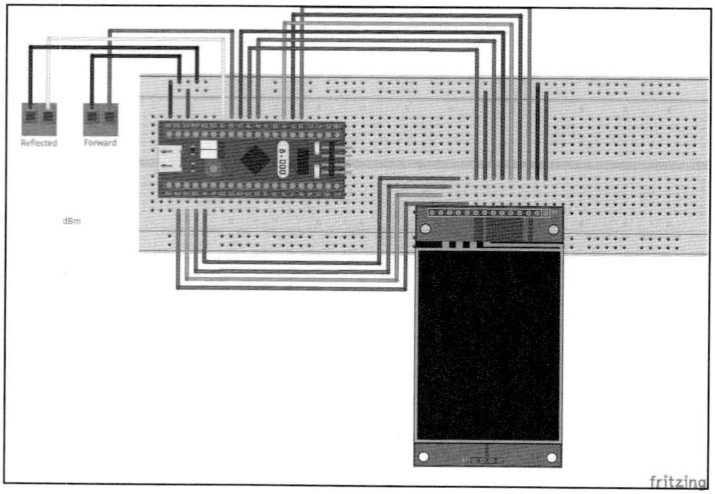

Figure 7.41 - Project Configuration

Project Build

Once you have copied the source repository to your local Arduino source folder, you are ready to build the project. Open the .ino file in the Arduino IDE and use the CTRL-U shortcut to compile and upload the project to our STM32 board.

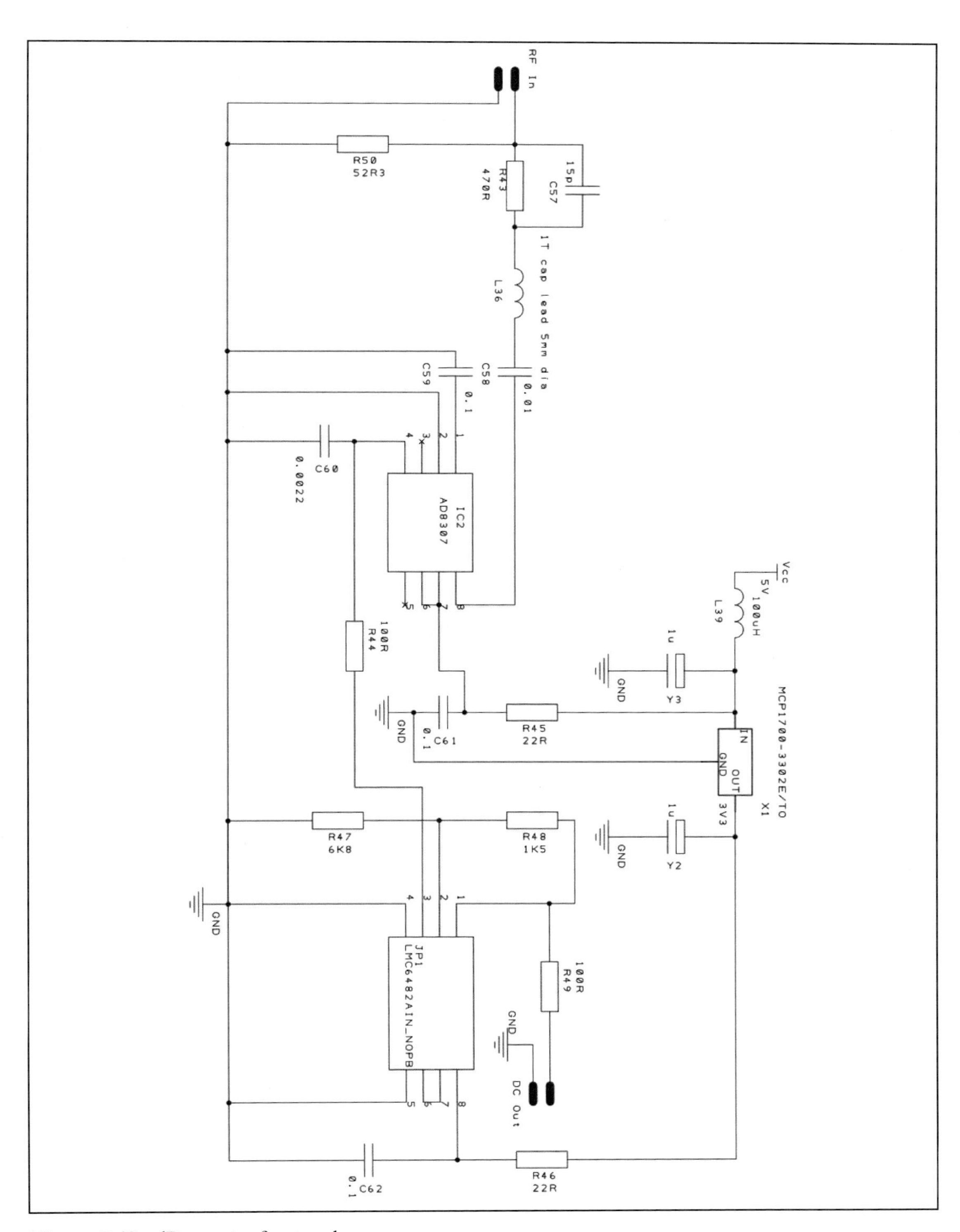

Figure 7.42 - dBm meter front end

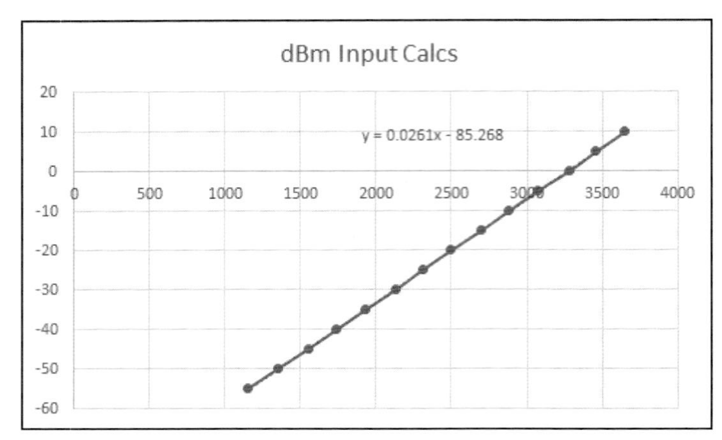

Figure 7.43 - dBm input stage

Figure 7.44 - dBm meter front end

Adding dBm direct reading

To add the direct reading power functionality to our project we will need a further front end to provide a 50 Ω Impedance to the outside world.

My experimental design is presented in **Figure 7.42**.

In this schematic the AD8750 is powered off a 5V rail and then uses a rail-to-rail op-amp powered at 3V3 to generate the input to our ADC. This maximises the voltage swing of the output to suit the MCU and ensures as much of the ADC resolution is used as possible.

In the same way as last time, I generated a range of input values using a signal generator and then plotted the resultant ADC values in Excel. Once again, I added a trendline and showed the associated equation to generate the algorithm to interpolate any ADC value back to a dBm reading. This graph is shown in **Figure 7.43**.

My poorly constructed version of this dBm meter front end is shown in **Figure 7.44**.

Software Changes

We now need to introduce the concept of two different modes to our power meter. We will have a forward/reflected mode where the unit operates as an SWR meter, and a second mode where it operates as a standalone power meter.

Project Configuration

The wiring diagram for the two-mode power meter is shown in **Figure 7.45**.

The new software will initialise in the SWR meter mode where forward, re-

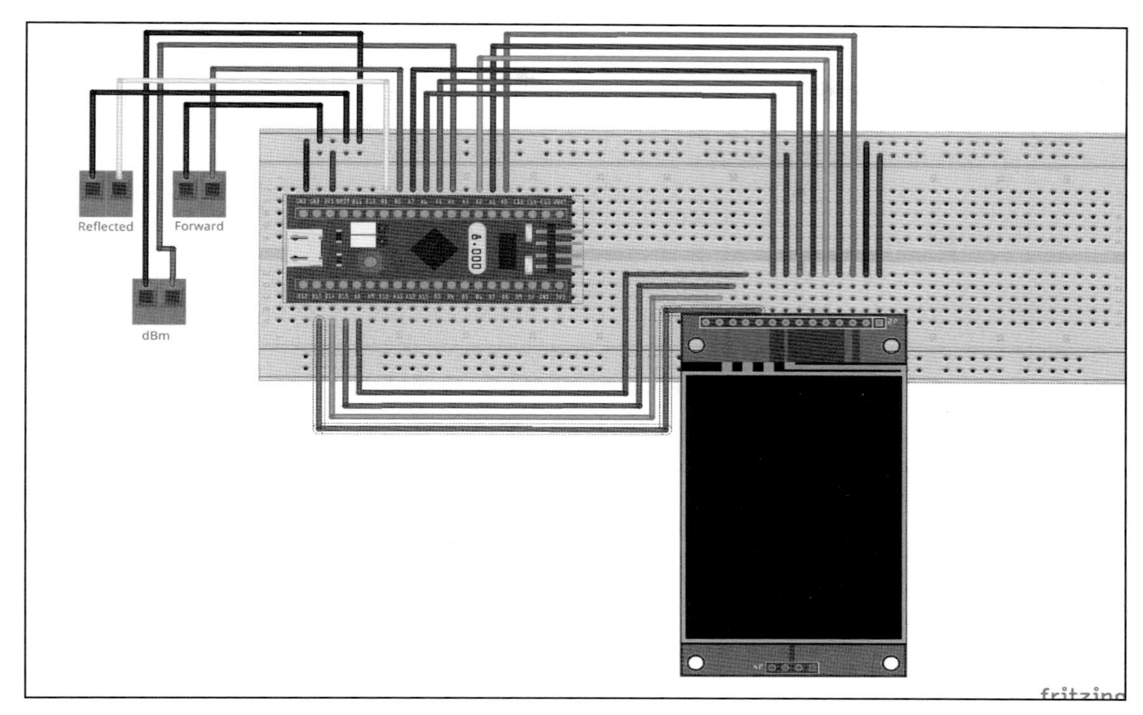

Figure 7.45 - Two Mode Power Meter Wiring

flected and SWR are displayed. A touch of the screen with change the mode to the new power meter mode.

In this new mode the power input as calculated in dBm is displayed as fed through the new front end into the dBm connector on the fritzing diagram above.

Project Support Files

All of the source code and supporting files for the project are contained in the Books extra page in the directory: Arduino\PowerMeterdBm

Figure 7.46 - Authors dBm/Power Meter

Project Build

Once you have copied the source repository to your local Arduino source folder, you are ready to build the project. Open the .ino file in the Arduino IDE and use the CTRL-U shortcut to compile and upload the project to our STM32 board.

The author's finished dBm/power meter is shown in **Figure 7.46** measuring an external -12dBm signal.

Things to try

1. DJ0ABR has published a design for a directional coupler which includes theAD8307s on the same board as the toroids – this is well worth a look and could be easily used in this project instead of my poorly constructed version.
2. What happens if you power on the final software with pin A5 grounded?

8

STM32CubeIDE

Shopping List

This section of the book uses the following components (Please refer to the hardware list file on the Books extra page for this book to obtain up-to-date URLs for where to buy these items):

- STLINK V2 dongle

Introduction

To date, we have utilised the Arduino IDE (Integrated Development Environment) for our projects. This has been customised for use with the STM32 processor board we selected and is using a core that provides support for the processor family. A great deal of the MCU configuration and initialisation is hidden from us in this environment and some features of the MCU are not accessible to us.

Our next projects will utilise the MCUs built in RTC (Real Time Clock) and Timers. To facilitate this, we will use the manufacturers IDE.

We are going to gain a better understanding of the software needed to initialise and configure the MCU and also of the overall architecture. The IDE will automatically generate a great deal of the code for us as a result of configuration actions in the GUI and throughout we will be using two manufacturer provided libraries of code:

- HAL the STMicroelectronics provided Hardware Abstraction Layer – a very comprehensive set of routines provided to access all of the peripherals and their configuration.
- CMSIS (pronounced Kim Sys) The ARM provided Cortex Micro-

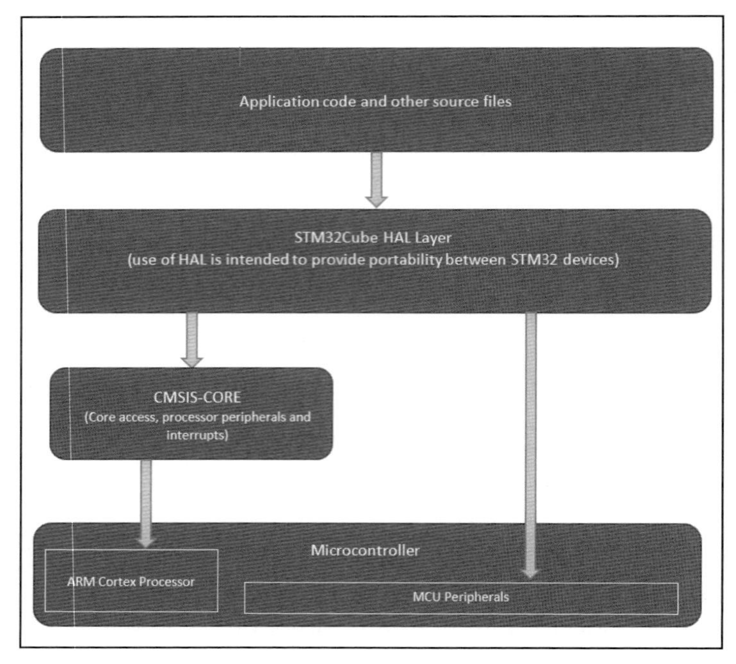

Figure 8.1 - STM32 Software Architecture Overview

controller Software Interface Standard – a very comprehensive set of routines provided to access all the Cortex processor components.

Remember that our STM32 MCU comprises an ARM processor plus STM peripherals, we need two sets of libraries to access all of its functions. This concept is illustrated in **Figure 8.1**.

IDE Installation

We need to install the STM32CubeIDE from STMicroelectronics. At the time of writing this was available at: https://www.st.com/en/development-tools/stm32cubeide.html

Download the appropriate version for your operating

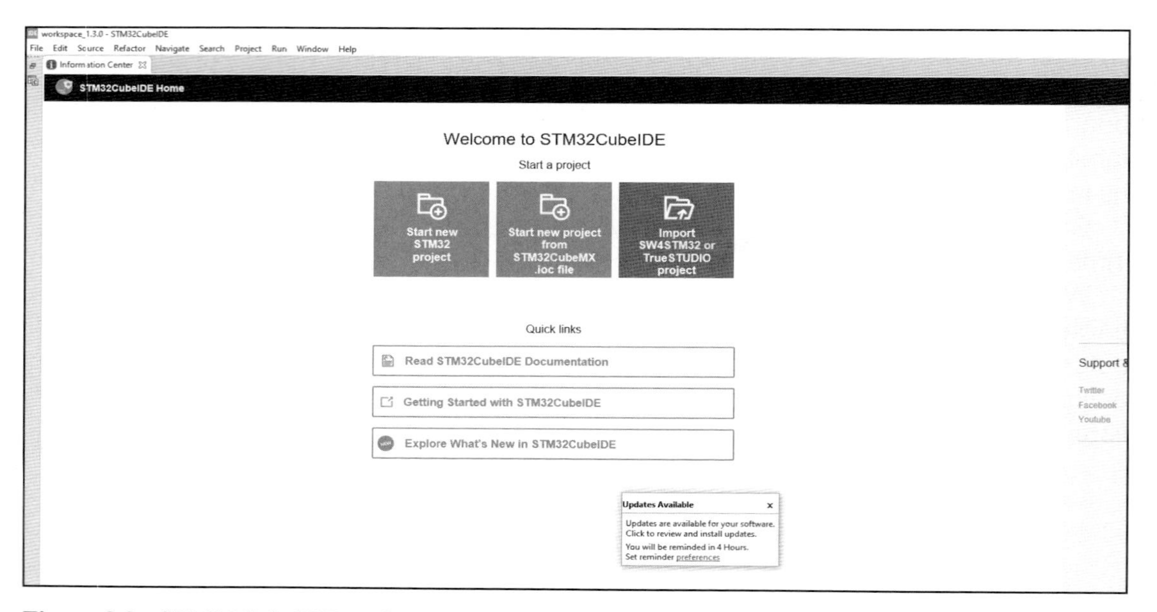

Figure 8.2 - STM32CubeIDE on first execution

system and run the executable to complete the installation. Follow the on-screen prompts and agree to install any device drivers you may be asked about.

Once installation is complete, we can run the IDE for the first time, you should be prompted to set the location of the workspace (I use the default), then you should see the same as **Figure 8.2**.

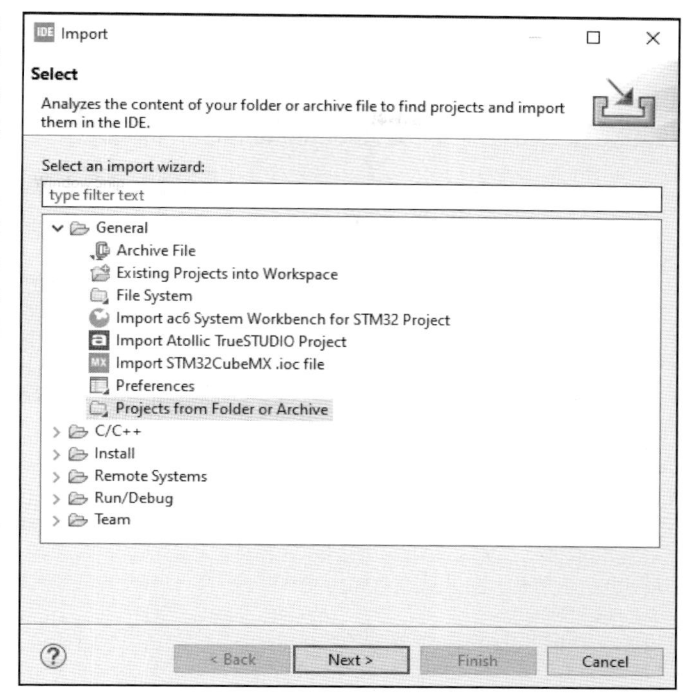

Figure 8.3 - Import from FileSystem option

Remembering Blinky

When we installed the Arduino IDE, we ran the embedded programmer's version of 'Hello World', a program to flash an LED on and off. We will do the same in our STM32CubeIDE, although the process will be more complicated as we need to undertake all the MCU configuration and initialisation that was automatically done for us by the Arduino IDE.

Note: *Each of the projects built for STM32CubeIDE is included in the Books extra page in the directory 'STM32CubeIDE Workspace'. You can import them all into the IDE using one action by using the File menu, Import, General and then selecting "Projects from folder or archive". Follow the on-screen instructions to completion. The import screen is show in* **Figure 8.3**.

Step 1 – New Project

Select from the File menu, new -> STM32 project.

This will start the Target Selector which may download files on its first execution. The STM32CubeIDE uses the concept of install on demand, application components are not downloaded to your local PC until they are needed. This means that each time you use a new processor or board type, the IDE will download the support files.

Step 2 - Processor Selection

Once the target selection window is open, make sure the MCU/MPU Selector tab is active and then type into the search box our processor type, which

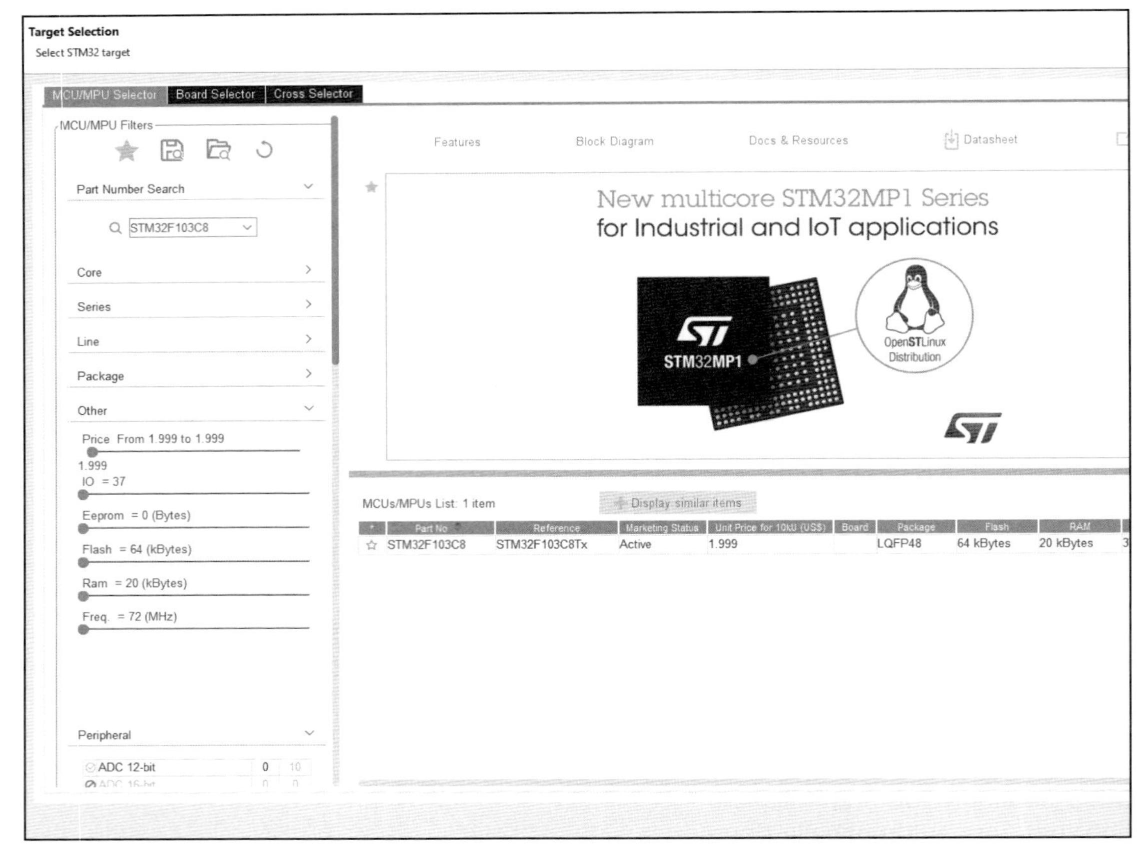

Figure 8.4 - STM32F103C8 Target Selection

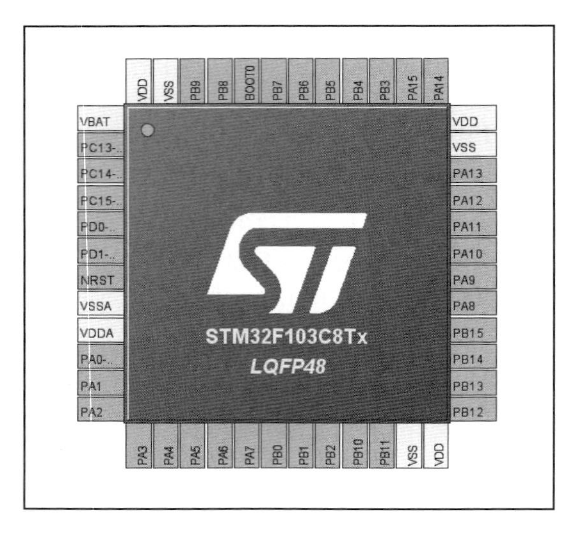

Figure 8.5 - MCU view in STM32CubeIDE

is: STM32F103C8. This should then show you a list of one processor to select from as shown in **Figure 8.4**.

Select the listed MCU and click next.

Step 3 – Project Name

We now need to give our project a name; I would suggest that you start the project name with the MCU type, so in this case we will call the project STM32F103C8-Blinky. Leave the Target Language, Target Binary Type and Targeted Project Type at their defaults and click finish.

You should now be presented with a graphical view of our MCU as shown in **Figure 8.5**.

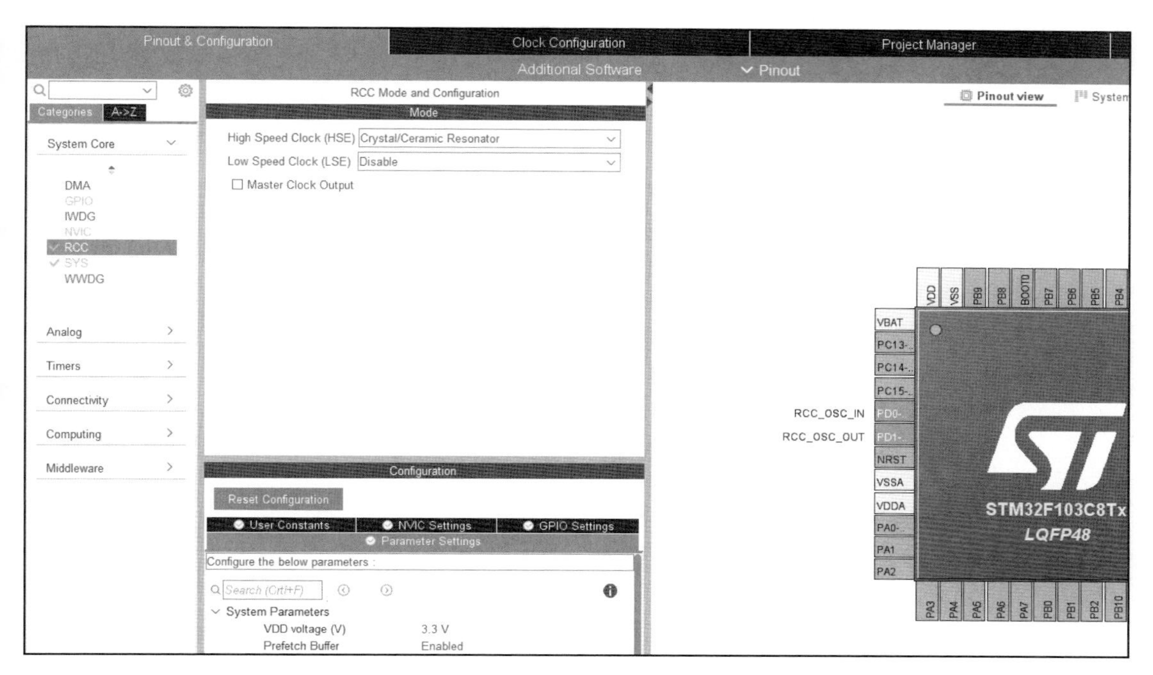

Figure 8.6 - HSE Selection

Step 4 – Clock Configuration

This view that we now have is designed to allow us to configure our MCU as required for our application hardware. As designers we can select from many different options, but in our example, we need to tell the IDE the configuration of our Blue Pill board.

Click on the left hand side of the screen and expand "System Core" then select "RCC".

You will now be able to select from drop down options for both the high-speed external (HSE) and low-speed external (LSE) clocks. Change the HSE selection to be "Crystal/Ceramic Resonator" as shown in **Figure 8.6**.

Now click on the Clock Configuration Tab and you should see the view shown in **Figure 8.7**.

We will now configure the clock to run using the High-Speed External (HSE) clock on our board which is an 8MHz crystal, then use the internal PLL to achieve a clock speed of 72MHz.

Change the radio button in the PLL source Mux to be the HSE, change the System Clock Mux to be PLLCLK and then enter a clock speed of 72 (MHz) into the HCLK box. Your clock settings should now look like **Figure 8.8**.

Step 5 - Debug Settings

Click on the Pinout & Configuration tab, then on the left-hand side under "System Core" select "SYS".

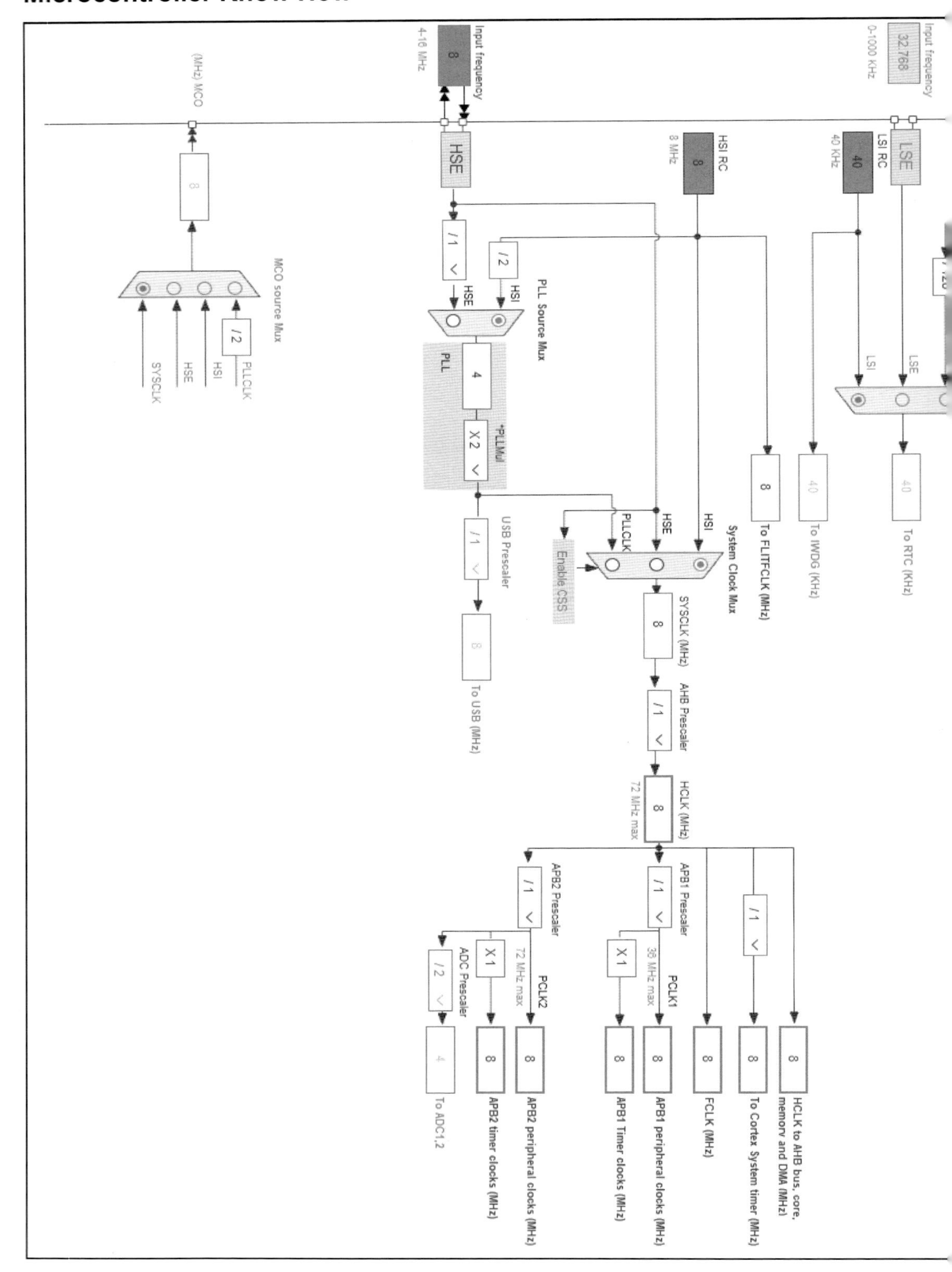

Figure 8.7 - Clock Configuration Tab

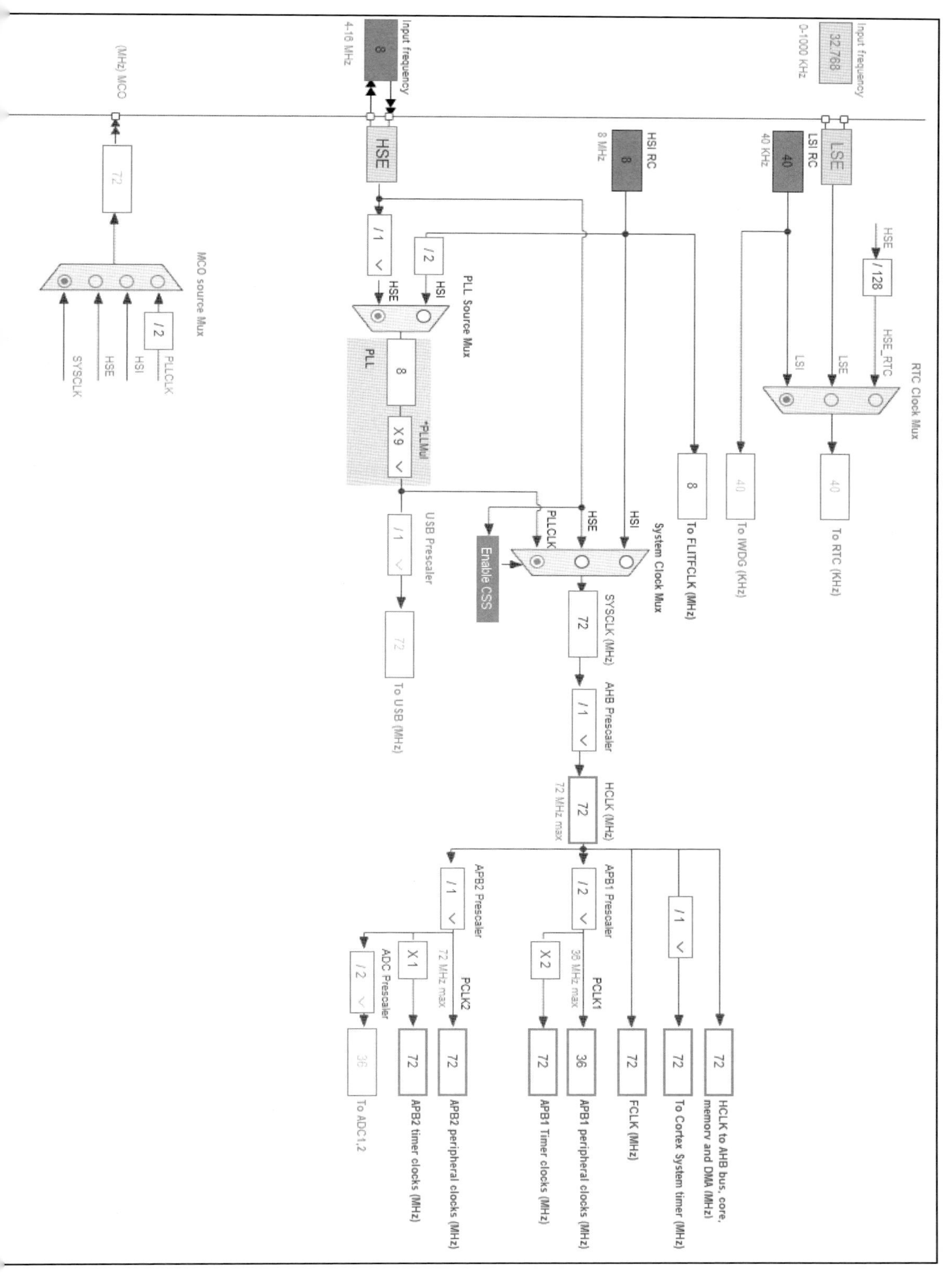

Figure 8.8 - Clock Configuration HSE, PLL 72MHz

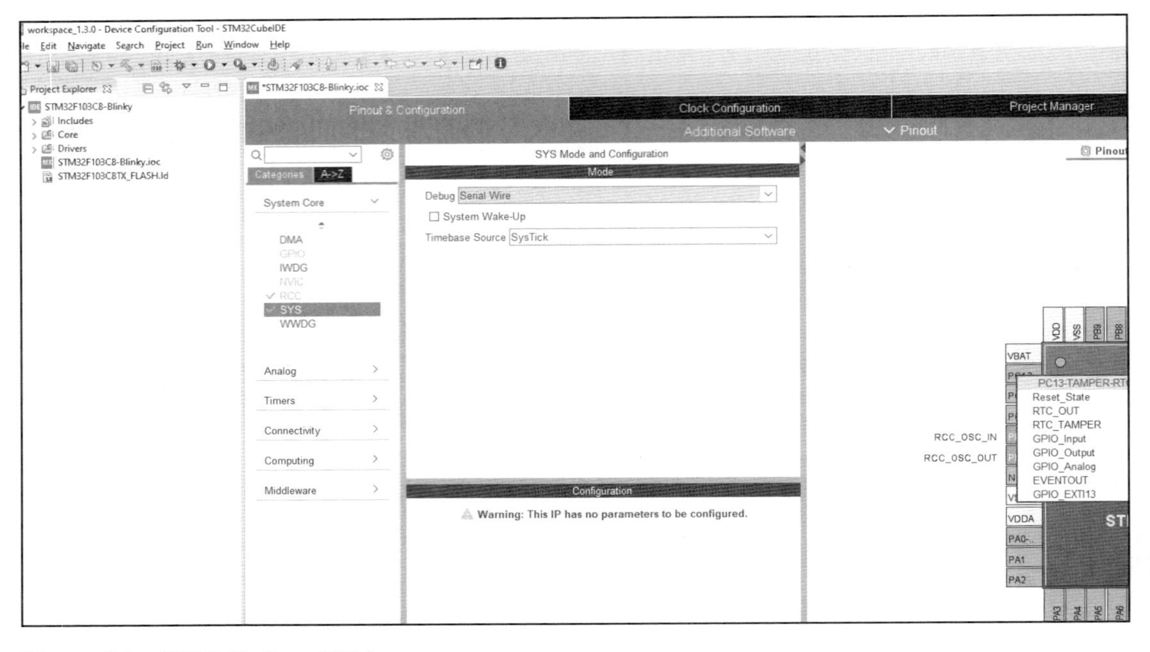

Figure 8.9 - GPIO Options PC13

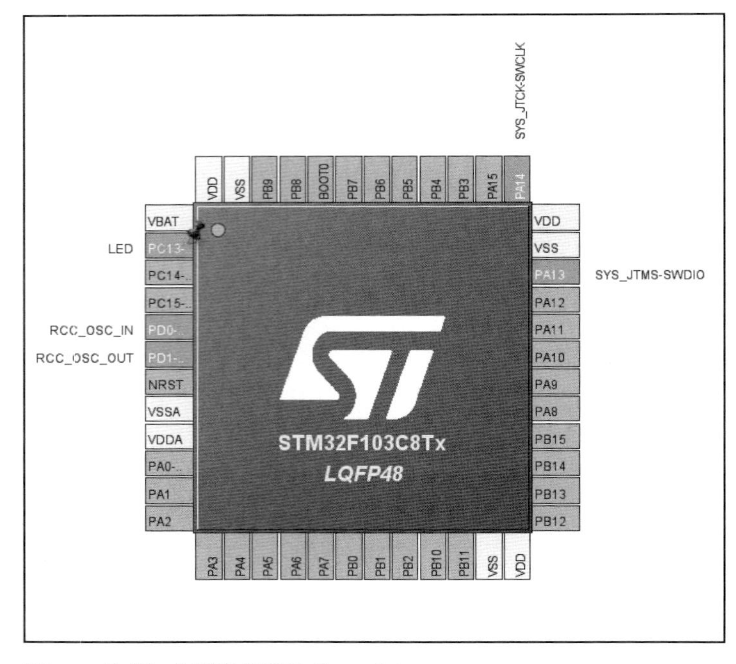

Figure 8.10 - MCU GPIO Complete

From the drop-down debug options select "Serial Wire".

Step 6 - Configure LED

Our Blue Pill board has the on-board LED connected to pin PC13 of the MCU. In the graphical illustration of our MCU, click on the pin in the diagram as shown in **Figure 8.9**.

What you can now see are all the possible configurations of Pin PC13, in our board this is physically wired to the on-board LED so we need to select GPIO_Output to configure the pin as a digital output.

Once you have made the selection, right click on the pin

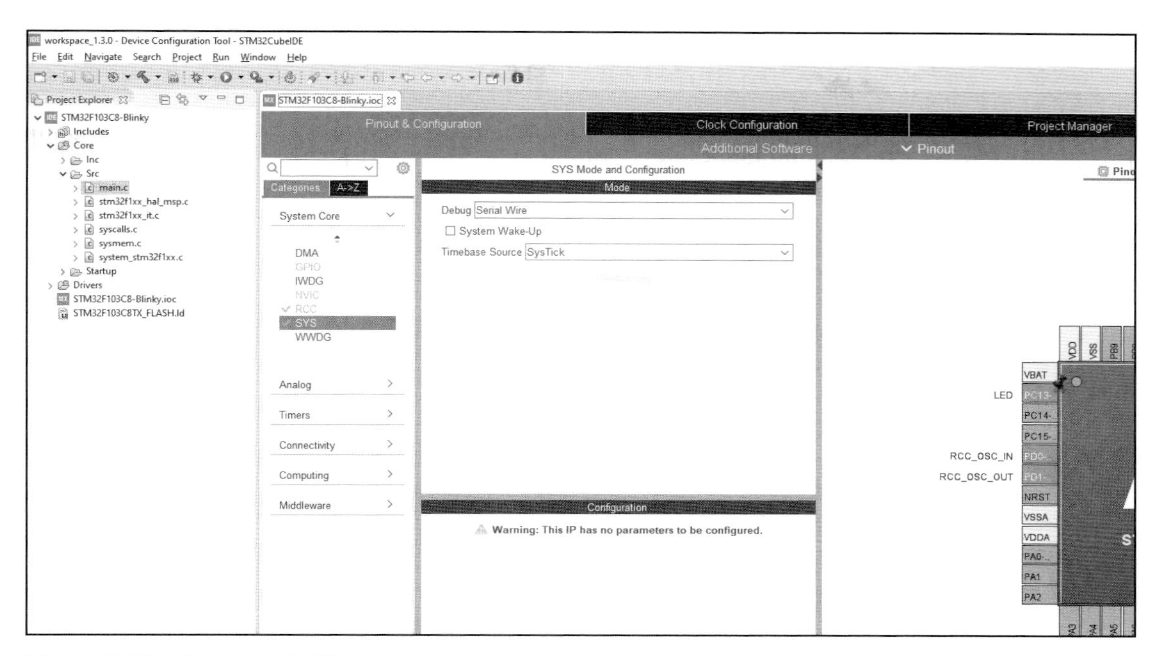

Figure 8.11 - Source code files

and select “Enter User Label” and enter the text “LED”. Your graphical representation of the MCU should now look like **Figure 8.10**.

Step 7 - Code Generation

We now use the IDE to generate all the configuration and initialisation code associated with the setup of the MCU we have undertaken. To do this select the file menu and save – the IDE will ask you if you wish to generate code. Answer yes to the prompt to get the IDE to generate the code for us.

Step 8 - Flashing the LED

So far, we have configured the IDE to work with our Blue Pill board. We configured the external 8MHz crystal as the clock, used the PLL to achieve a processor speed of 72MHz and set up the GPIO pin PC13 as an output.

We now need to write the functionality we require in our application, flashing the LED. On the left-hand side of the IDE you need to expand the core section of the hierarchy, then src to reveal the source code files that have been generated. This is shown in **Figure 8.11**.

Double click on the file “main.c” and it will open into an editor as shown in **Figure 8.12**.

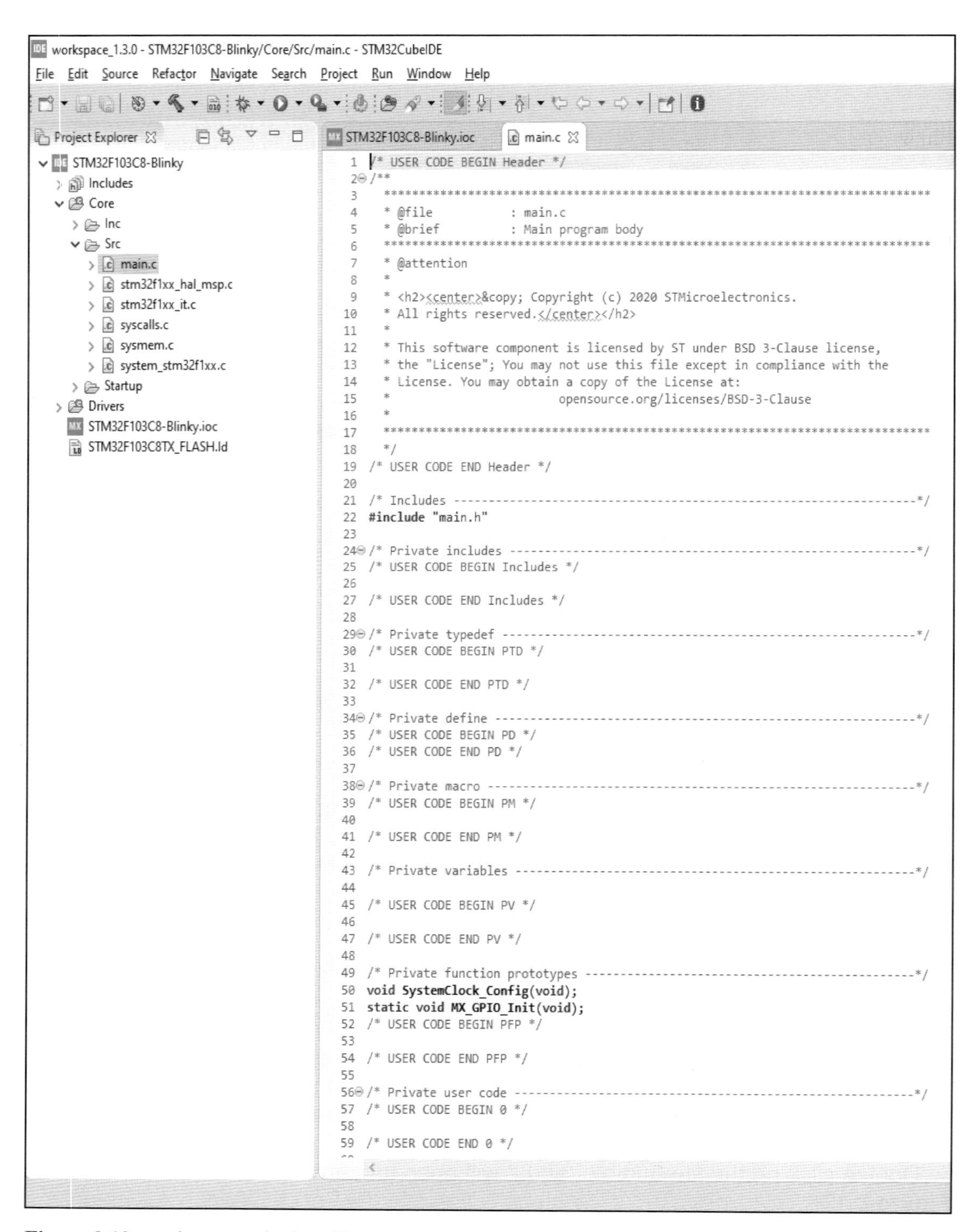

Figure 8.12 - main.c open in the editor

workspace_1.3.0 - STM32F103C8-Blinky/Core/Src/main.c - STM32CubeIDE

File Edit Source Refactor Navigate Search Project Run Window Help

Project Explorer

- STM32F103C8-Blinky
  - Binaries
  - Includes
  - Core
    - Inc
    - Src
      - main.c
      - stm32f1xx_hal_msp.c
      - stm32f1xx_it.c
      - syscalls.c
      - sysmem.c
      - system_stm32f1xx.c
    - Startup
  - Drivers
  - Debug
  - STM32F103C8-Blinky.ioc
  - STM32F103C8TX_FLASH.ld

STM32F103C8-Blinky.ioc | main.c

```
61 /**
62   * @brief  The application entry point.
63   * @retval int
64   */
65 int main(void)
66 {
67   /* USER CODE BEGIN 1 */
68
69   /* USER CODE END 1 */
70
71   /* MCU Configuration--------------------------------------------------------*/
72
73   /* Reset of all peripherals, Initializes the Flash interface and the Systick. */
74   HAL_Init();
75
76   /* USER CODE BEGIN Init */
77
78   /* USER CODE END Init */
79
80   /* Configure the system clock */
81   SystemClock_Config();
82
83   /* USER CODE BEGIN SysInit */
84
85   /* USER CODE END SysInit */
86
87   /* Initialize all configured peripherals */
88   MX_GPIO_Init();
89   /* USER CODE BEGIN 2 */
90
91   /* USER CODE END 2 */
92
93   /* Infinite loop */
94   /* USER CODE BEGIN WHILE */
95   while (1)
96   {
97       HAL_GPIO_TogglePin(LED_GPIO_Port, LED_Pin);
98       HAL_Delay (1000);
99     /* USER CODE END WHILE */
100
101     /* USER CODE BEGIN 3 */
102   }
103   /* USER CODE END 3 */
104 }
105
```

Console

CDT Build Console [STM32F103C8-Blinky]

```
Finished building: STM32F103C8-Blinky.bin
Finished building: STM32F103C8-Blinky.list


08:46:06 Build Finished. 0 errors, 0 warnings. (took 4s.312ms)
```

Figure 8.13 - Build Complete

We are going to scroll down in the main.c file until we find this section:

```
/* Infinite loop */
/* USER CODE BEGIN WHILE */
while (1)
{
  /* USER CODE END WHILE */
```

And we are going to change this code to look like this:

```
/* Infinite loop */
/* USER CODE BEGIN WHILE */
while (1)
{
          HAL_GPIO_TogglePin(LED_GPIO_Port, LED_
          Pin);
          HAL_Delay (1000);
  /* USER CODE END WHILE */
```

Once you have modified the code, we will build the project. You can select this by clicking on the hammer icon in the main window. Once the build is complete your screen should look like **Figure 8.13**.

Step 9 - Connecting to the Target

We are going to use an inexpensive STLINK V2 dongle to connect our IDE to our blue pill board as it provides debugging capabilities that out previous FTDI serial interface does not.

Wire the STLINK V2 dongle to the STM32 Blue Pill board as shown in **Figure 8.14**. Next connect the STLINK V2 dongle to a USB port of the PC and it should install a device and appear in the Device Manager under Universal Serial Bus Devices and be called "STM32 STLink" as shown in **Figure 8.15**.

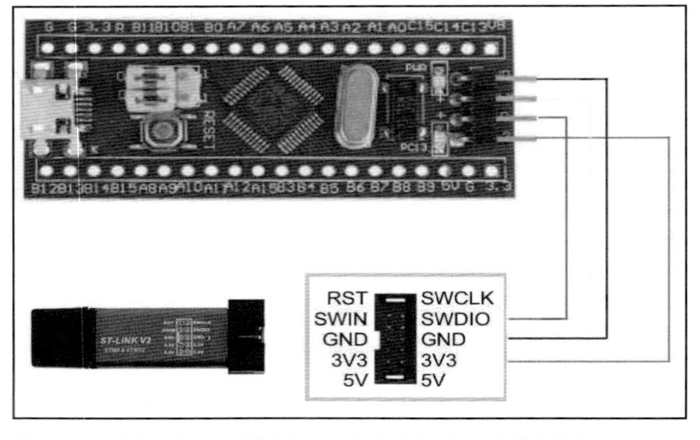

Figure 8.14 - STLINK V2 to STM32 Blue Pill Wiring

We will now download our project to the target using the STM32CubeIDE. To complete this step, click on the play button icon on the toolbar, you will be prompted to confirm the configuration of the build, just click to continue without making any changes.

If you encounter a failure to connect to the board, move the boot0 jumper on the blue pill board to the 1 position, press the reset button on the target board and try again. Once successfully loaded replace the boot0 jumper back to the 0 position.

You will now have the LED on the blue pill board flashing at 1 second intervals; congratulations you just completed your first STM32CubeIDE project.

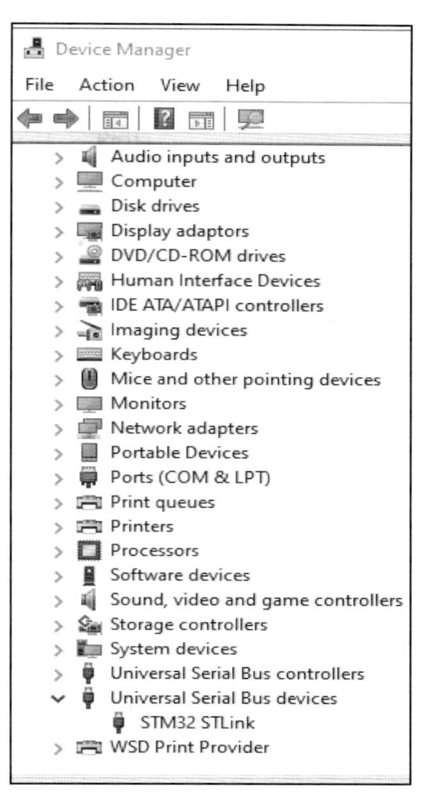

Figure 8.15 - STLink Device Manager Entry

Conclusions

Whilst it may seem far more laborious to create the blinky example using the STM32CubeIDE rather than the Arduino version we did previously, I hope you can appreciate that we were in control of many more of the configuration and initialisation choices for the MCU. We have selected the processor speed, the clock type being used as the HSE, the configuration of the MCU pins (in our case PC13), we have even labelled the pin for use within the code. To fully utilise the peripherals within the MCU we need to operate at this level, the Arduino IDE takes us further away from the details and makes many choices on our behalf. If we were designing our own embedded systems, I hope you can see that the choices made by the Arduino IDE may not be correct for our circumstances.

Things to Try

1. Change the Clock Configuration to use the internal HSI (High Speed Internal) Clock and change the overall clock speed set in HCLK. Re-generate the code and see what difference it makes.
2. Try and research the advantages of using the HSE over the HSI clock.
3. Change the frequency of the LED's flashing.

Notes on STM32CubeIDE

Notes on Code Auto-Generation

If you worked through the blinky example, you will have seen that a large amount of the code we used was auto-generated by the IDE. At any point you can change the configuration of the MCU and re-generate the code to reflect your changes.

Please note that it is therefore essential that any user written code is inserted within the correct parts of the main.c (or other) files. Each section of the file contains a start and end comment for user code sections:

```
/* Infinite loop */
/* USER CODE BEGIN WHILE */
while (1)
{
        ← put user code in here between BEGIN and
        END comments
  /* USER CODE END WHILE */

  /* USER CODE BEGIN 3 */
}
/* USER CODE END 3 */

/* USER CODE BEGIN 4 */
      ← put user code in here between BEGIN and END
      comments
/* USER CODE END 4 */
```

Failure to put the user code in the appropriate sections within the code files will result in it being deleted when auto-generation is run.

There are separate sections in the files for user code for initialisation, main loop instructions, function/procedure definitions and contents.

A word about scheduling

Up to now, all our examples have used delay statements to wait for time to pass. This is generally bad practice as it is known as 'blocking code' i.e. the processor cannot do anything else whilst the delay statement is executing and therefore is 'blocked'. It is a good idea to get used to the idea of scheduling events within your developed code and an example has been provided in the Books extra page within the folder STM-32CubeIDE Workspace called STM32F103C8-Multiple-Blinky.

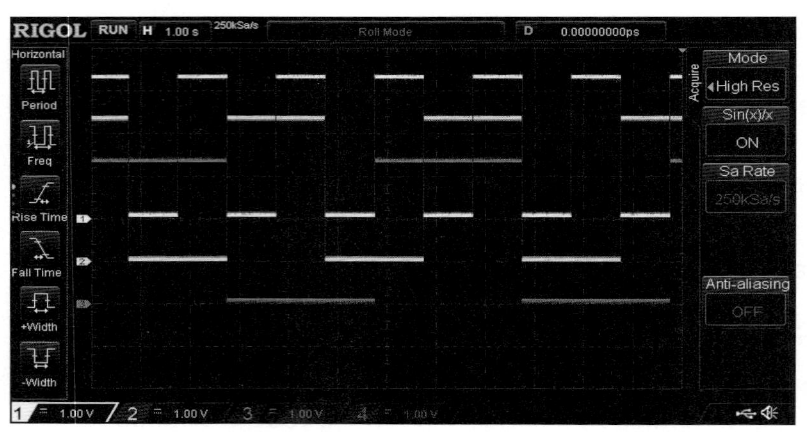

Figure 8.16 - Multiple Blinky Output

Please have a look at this code and understand how the events are timed and executed. The code uses the concept of a system tick – which is a 1ms counter. The code simply calculates elapsed time since the last event and then checks to see if it is time to do something. If it is, the code acts and stores the current time for the next time round. In the example, you will see that there are three separate timed events repeatedly occurring. This is not possible using the delay function and is also considered good practice as the processor is not being blocked to achieve the timing. The output from this example is shown in **Figure 8.16** and illustrates the three output pins PC13, PC14, PC15 toggling high and low at 1, 2 and 3 second intervals respectively.

Things to Try

1. Get the multiple-Blinky application running on a breadboard with three LEDs and appropriate series resistors.
2. Change the event frequency of each output.

A word about Debugging

The Arduino IDE provides no real mechanism for debugging code. At best you must use the serial interface and place multiple print statements inside the code to output information as it runs.

In the STM32CubeIDE environment combined with the STLINK V2 dongle, we have an extremely powerful debug facility. The debugger allows you to step through your code as it executes on the target and see the results of individual

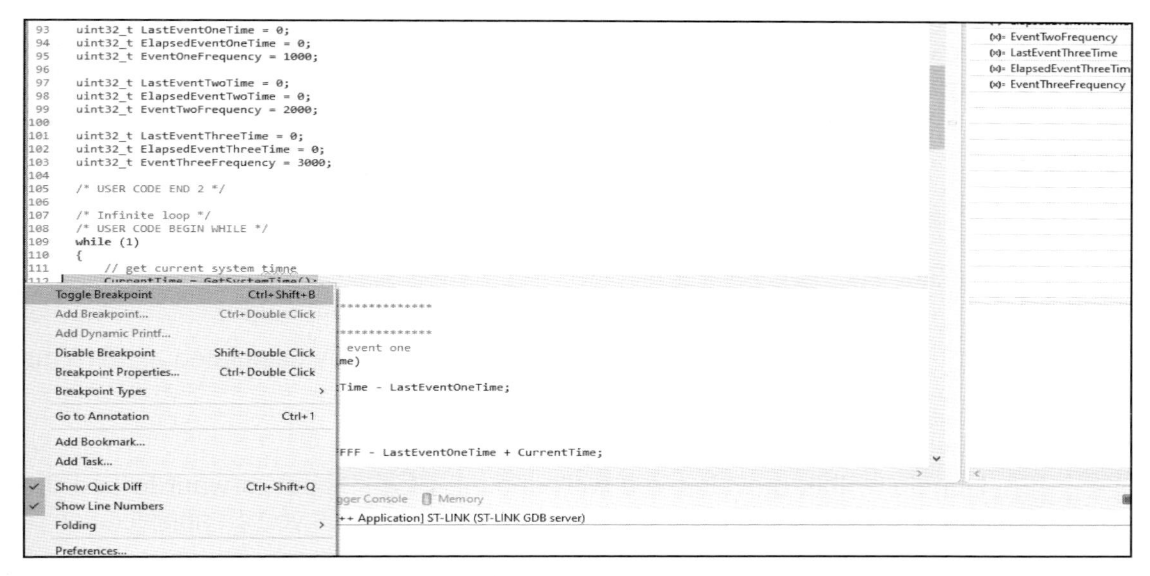

Figure 8.17 - Toggle a Breakpoint in the Debugger

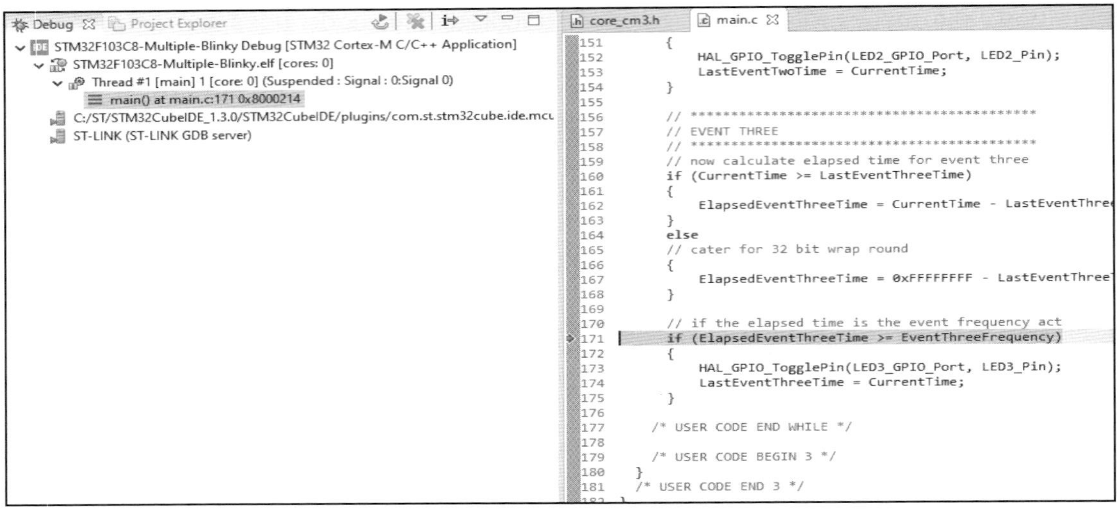

```
Debug  Project Explorer
STM32F103C8-Multiple-Blinky Debug [STM32 Cortex-M C/C++ Application]
  STM32F103C8-Multiple-Blinky.elf [cores: 0]
    Thread #1 [main] 1 [core: 0] (Suspended : Signal : 0:Signal 0)
      main() at main.c:171 0x8000214
  C:/ST/STM32CubeIDE_1.3.0/STM32CubeIDE/plugins/com.st.stm32cube.ide.mcu
  ST-LINK (ST-LINK GDB server)
```

```
core_cm3.h   main.c
151       {
152          HAL_GPIO_TogglePin(LED2_GPIO_Port, LED2_Pin);
153          LastEventTwoTime = CurrentTime;
154       }
155
156       // ********************************************
157       // EVENT THREE
158       // ********************************************
159       // now calculate elapsed time for event three
160       if (CurrentTime >= LastEventThreeTime)
161       {
162          ElapsedEventThreeTime = CurrentTime - LastEventThre
163       }
164       else
165       // cater for 32 bit wrap round
166       {
167          ElapsedEventThreeTime = 0xFFFFFFFF - LastEventThree
168       }
169
170       // if the elapsed time is the event frequency act
171       if (ElapsedEventThreeTime >= EventThreeFrequency)
172       {
173          HAL_GPIO_TogglePin(LED3_GPIO_Port, LED3_Pin);
174          LastEventThreeTime = CurrentTime;
175       }
176
177     /* USER CODE END WHILE */
178
179     /* USER CODE BEGIN 3 */
180   }
181   /* USER CODE END 3 */
```

Figure 8.18 - Code Halted during Execution

statements as they are undertaken. This is an extremely powerful tool that would warrant an entire volume on the detailed functionality available to the user.

The debugger is accessed by using the bug icon, I would encourage you to spend some time researching the debugger functionality and getting to grips with its features. In **Figure 8.17** I have set a breakpoint inside the code of our 'Multiple-Blinky' application, the debugger has executed the code to this point and then halted. You can see on the right-hand side of the screen a list of all the variables in scope at this point in the code and highlighted in yellow are those that have changed.

We can then step through the code a line at a time using and if you look at **Figure 8.18** we can glean the following information:

- The value in Current Time is 4318
- The last event one time was 4000 (318 ms ago)
- The Elapsed event one time has correctly calculated at 318 ms
- The last event three time was 3000 (1319 ms ago)
- The Elapsed event three time has correctly calculated at 1318 ms
- Et cetera

Hopefully, this provides you with enough of a snippet to want to use the debugger when fault finding in your code and also gives you an appreciation of the powerful functionality available.

None of this is available in the Arduino IDE.

9

STM32CubeIDE Project
A GPS synced clock

Shopping List

This section of the book uses the following components (Please refer to the hardware list file on the Books extra page for this book to obtain up-to-date URLs for where to buy these items):

- GPS Module
- I2C LCD Display

Introduction

We are going to introduce some new concepts into our software through this project. We will use the built in RTC (Real Time Clock) within our MCU and we will also interface a GPS module using a serial data stream. To set our RTC we will parse the serial data to extract the time. We can then use the internal RTC time to display UK daylight saving adjusted time on an I2C interfaced LCD.

You can import the project from the RSGB book file repository, the STM-32CubeIDE project is contained in folder STM32CubeIDE Workspace and is called STM32F103C8-Blue-Pill-Clock.

Alternatively, you can go through the exercise of configuring the project yourself from the instructions below by using the IDE and then copy in the code files as required.

STM32CubeIDE Configuration

Configuration of the project will be like the Blinky example we completed previously with some additions:

- Configure the LSE (Low Speed External) clock to use the on-board 32.768 kHz crystal and configure this to be the source for our RTC (Real Time Clock). This is shown in **Figure 9.1** and **Figure 9.2**.
- Enable the RTC and calendar in the "Timers" section of the configuration and set the data format to be binary and the day/date/year to be a valid combination. This is shown in **Figure 9.3**.
- Enable USART3 for serial comms with our GPS and set the baud rate to be 9600 bits/s. This is shown in **Figure 9.4**.
- Enable interrupts for USART3 . This is shown in **Figure 9.5**.
- Configure USART1 for serial comms with our host PC and set the baud rate to be 115200 bits/s. This is shown in **Figure 9.6**.
- Enable the built in I2C interface to operate in the standard I2C mode and use this to interface to an LCD display. This is shown in **Figure 9.7**.
- Finally, we will force the IDE Code Generator to "generate peripheral initialisation as a pair of '.c/;h' files per peripheral" by selecting this option in the Project Manager tab under Code Generator. This is shown in **Figure 9.8**.

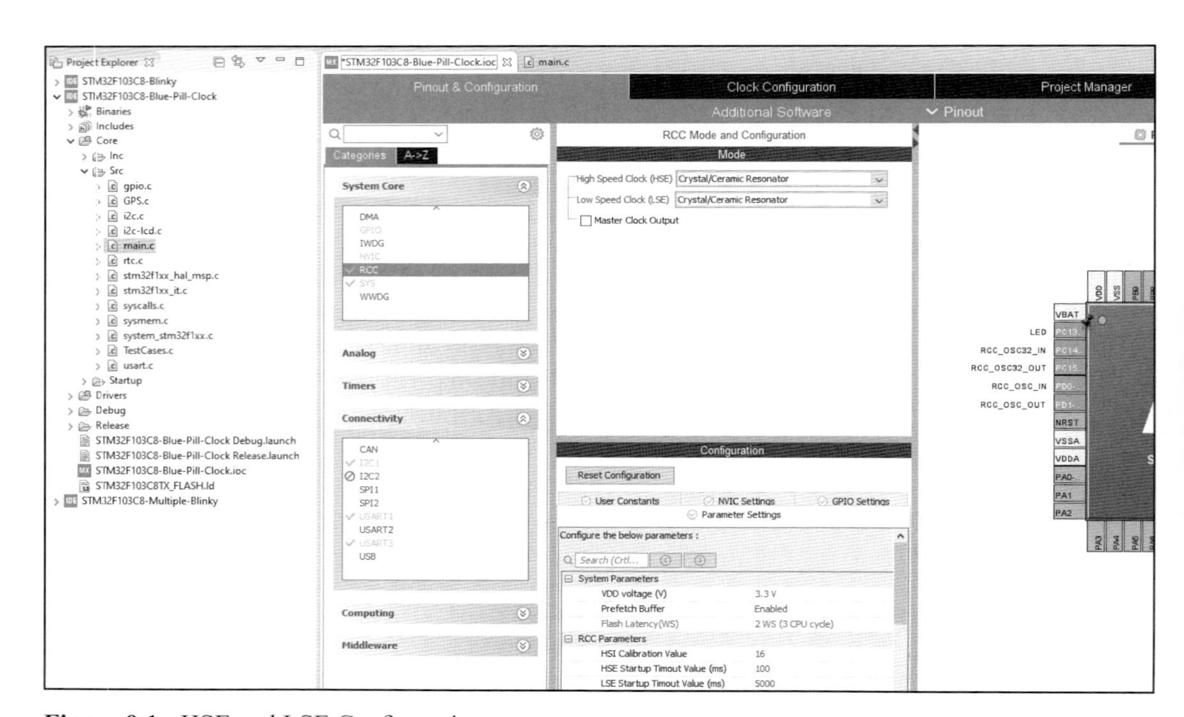

Figure 9.1 - HSE and LSE Configuration

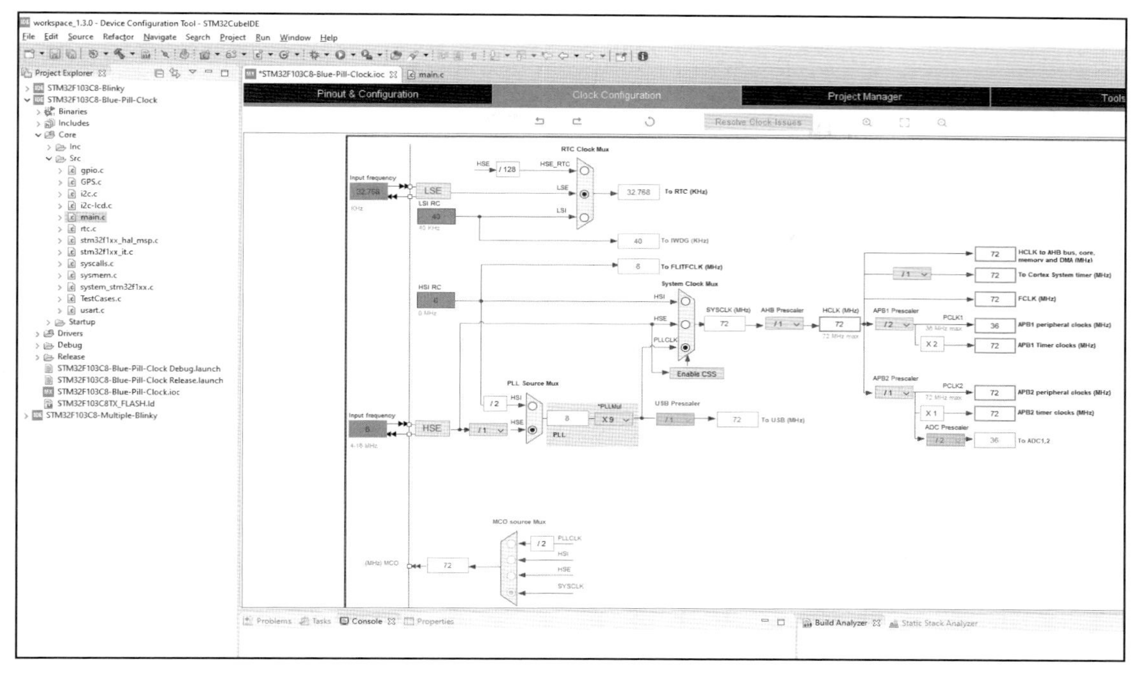

Figure 9.2 - Clock Configuration

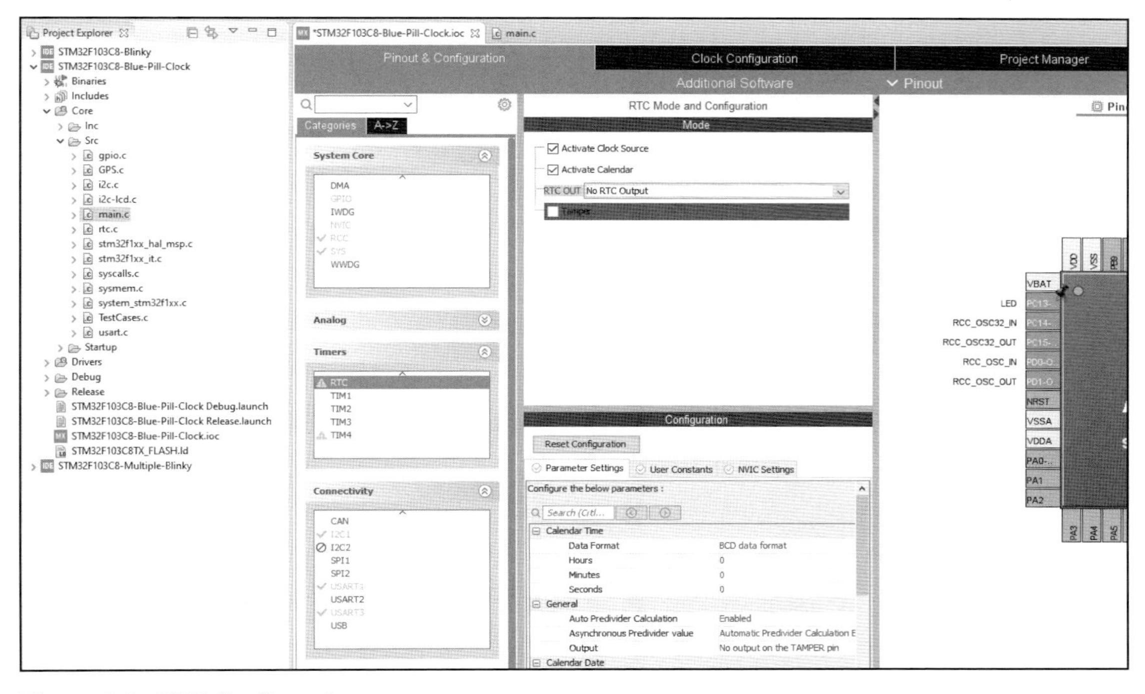

Figure 9.3 - RTC Configuration

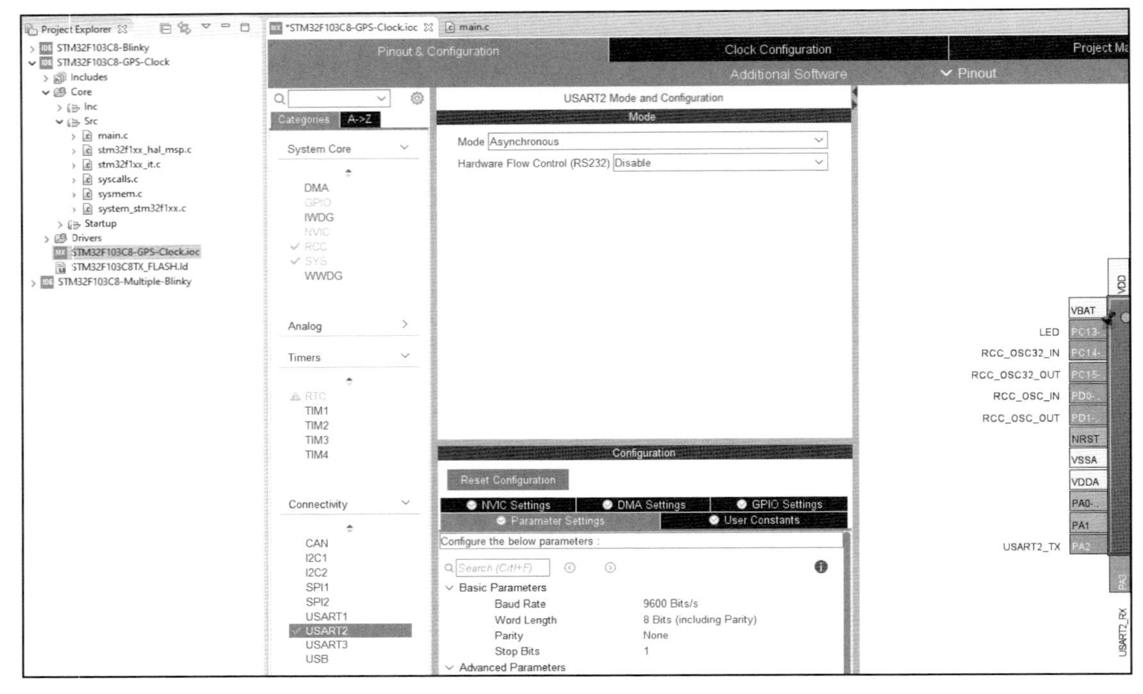

Figure 9.4 – USART3 Configuration

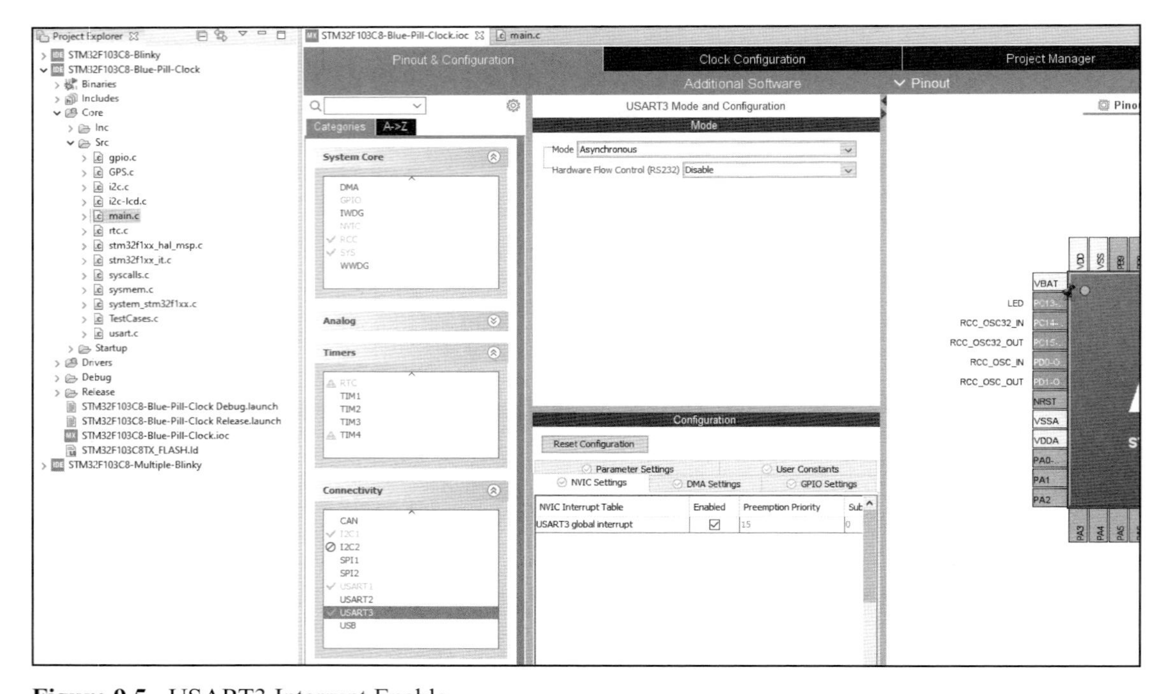

Figure 9.5 - USART3 Interrupt Enable

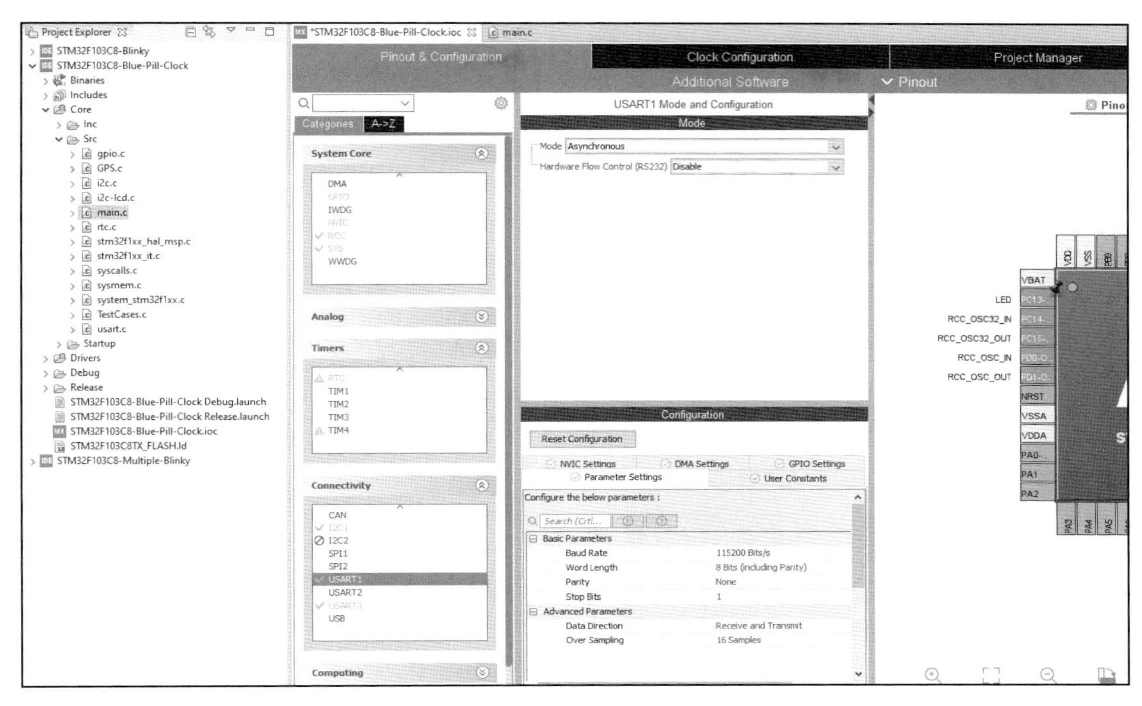

Figure 9.6 - USART1 Configuration

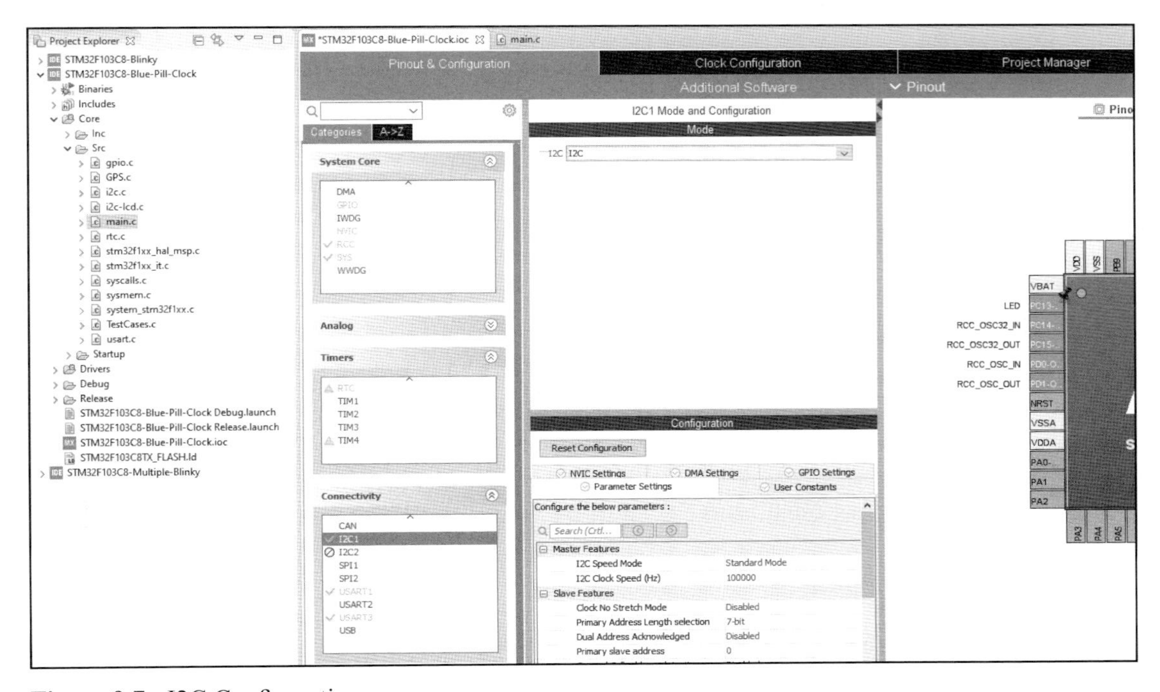

Figure 9.7 - I2C Configuration

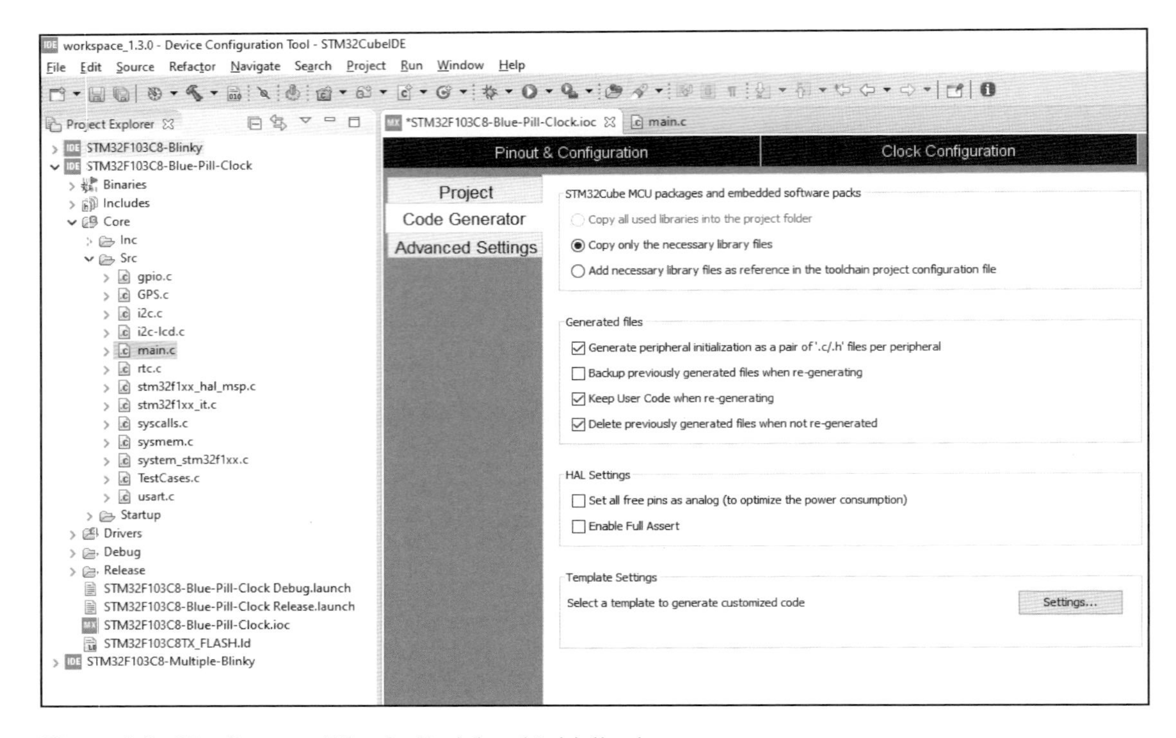

Figure 9.8 - Use Separate Files for Peripheral Initialisation

The complete list of the configuration steps needed:

- Create a new STM32Cube project using MCU STM32F103C8
- Section RCC – use HSE and LSE as Ceramic Resonator/Crystal
- Section SYS – use serial wire debug
- Section Timers – Enable RTC
    - Activate calendar
    - Select binary data format
    - Set calendar date to a correct day/date/year e.g. Monday / 5th October / 2020
- Set GPIO pin PC13 to be output and enter user label LED
- Section Connectivity – enable USART3 and set its speed to be 9600 baud
- Section Connectivity – enable USART1 and set its speed to be 115200 baud

- Section Connectivity – enable I2C1 in standard I2C mode
- Project Manager Tab
    - o Code Generator – select "generate peripheral initialisation as a pair of '.c/;h' files per peripheral"
- Clock Configuration Tab
    - o Select LSE as source for RTC
    - o Select HSE and PLL input with clock speed of 72MHz
- Generate code

A word about interrupts

Imagine you are sat watching television, cup of tea in hand when suddenly the doorbell rings. You would probably put down your tea, get up from your chair and go to answer the door. Your current activity has been interrupted and you immediately go and do something more important. Generally, these events are time critical – if you do not answer the door promptly, likelihood is that the visitor will leave.

For the first time in this project, we introduce the concept of interrupts and the concept above is an excellent example of what happens to an MCU when an interrupt occurs. The processor is interrupted from performing its current task to go and do something of a higher priority. When USART3, which is receiving data from our GPS, has data ready for processing, it generates an interrupt and we must provide the necessary code to handle that interrupt and process

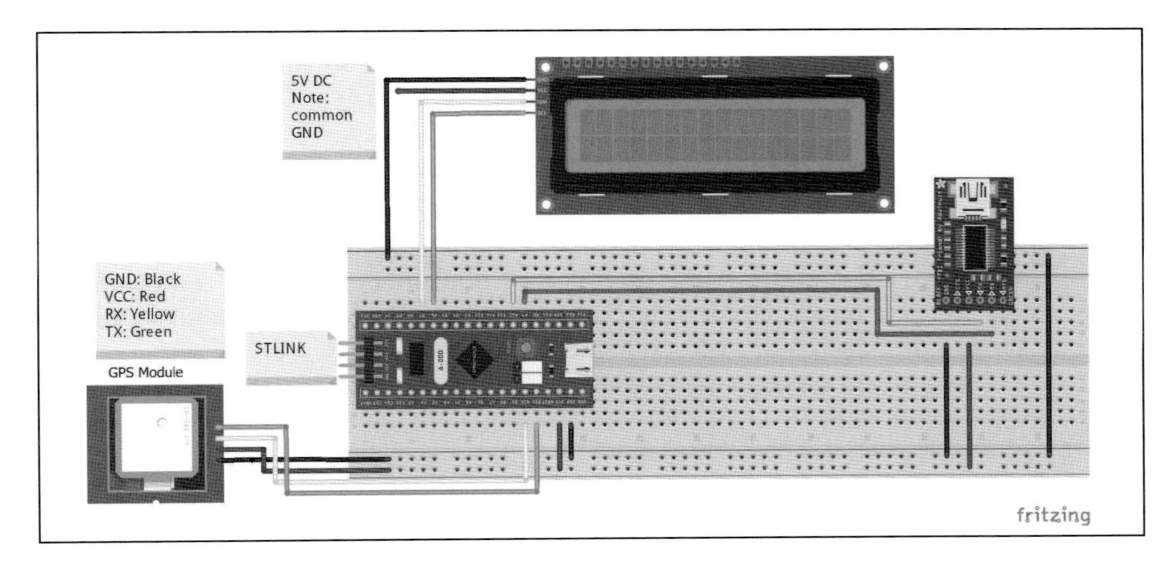

Figure 9.9 - GPS Clock Configuration

the data. Routines written to handle interrupts are usually called CallBacks.

This is a convenient way of dealing with busy data streams from devices like the GPS as it means we do not need to keep checking the data, but instead rely on the hardware to interrupt the software to tell us that action is needed.

The routine that has been developed for handling the GPS USART data is called GPS_CallBack and can be found in the GPS.c file in our project code.

Project Configuration

Wire up your project as shown in **Figure 9.9**. Note that we are re-using the FTDI USB to Serial adaptor from the initial bootloader programming task to act as the debug communications interface back to the PC through USART1.

There are many ways to provide debug information, this is included here just as an example of one such mechanism. You will find a variable declared in the code called SerialDebug, when this is set to TRUE, elsewhere in the code serial information is sent to the PC to provide us with debug information.

Project Build

Once you have imported the project from the Books extra page into your IDE, make sure that 'STM32F103C8-Blue-Pill-Clock' is selected in the Project Explorer, your STLINK is connected to the board as the programming interface and you are supplying the board with power, then click on the Play Icon to compile, download and execute the code on the target microcontroller.

Note that the STLINK module will provide the 3V3 supply to the electronics, the LCD however, requires a separate 5V supply. Be sure to keep them separate but connected to the common ground.

Once running you should initially see the screen display "Waiting for GPS" and then, once the GPS has locked to a satellite, you will see the date and time displayed.

A word about I2C

Our LCD module uses I2C (pronounced I squared C) sometimes called I2C (I two C) Inter-Integrated circuit serial communications for the MCU to talk to the LCD. Developed in 1982 by Phillips Semiconductors® it is a mechanism designed for short distance serial communications between low-speed peripheral ICs and processors or microcontrollers.

The protocol uses a serial clock line (SCL) and a serial data line (SDA) and has the concept of a slave address, meaning that multiple devices with different addresses can reside on the same bus.

In our example the address of the LCD interface is hex 4E and you will find this defined in the i2c-lcd.c file. Most I2C compatible hardware devices will have a default address, but usually include a mechanism of changing the address in

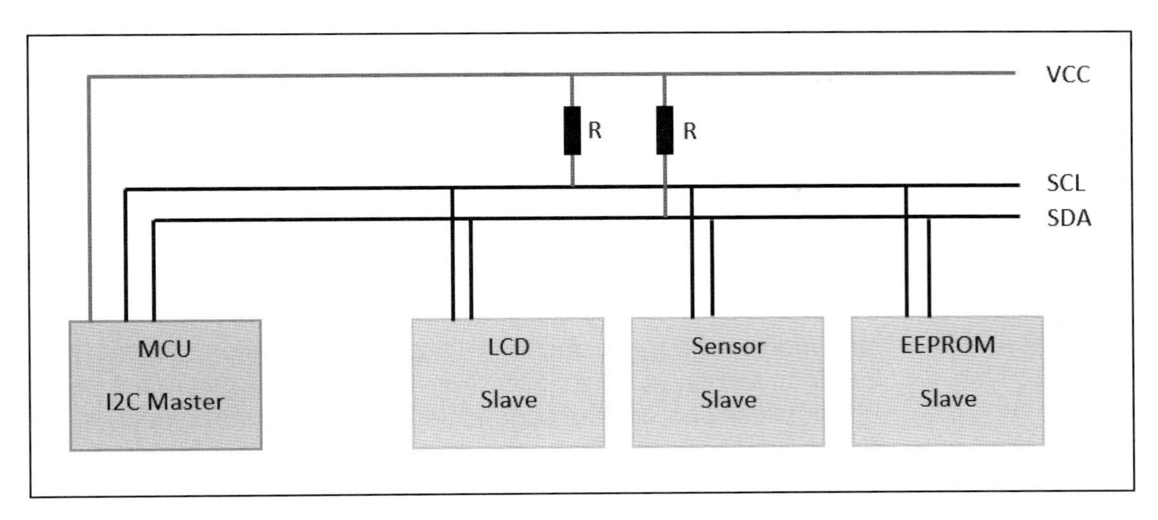

Figure 9.10 - I2C Overview

the event of a clash with another peripheral. In our example, there are spaces on the I2C LCD interface board for links labelled A0, A1, A2 which alter the device address.

Things to Try

1. The IDE has been configured to create separate files for each peripheral initialisation. The GPS and I2C_LCD interfaces are also included in separate files. See if you can understand why it has been done this way.
2. Take a good look at the code, there are several interesting routines included:

   a. The GPS parse routine is experimental and uses the GPRMC string from the GPS. Try and understand how this is working to extract the relevant information we need for our display.

   b. The Daylight-Saving calculation is exceptionally fiendish, most notably the routine that calculates the day of the week from any given date. You will find this in the main.c file and is called "DayofWeek". Do some research on this algorithm and see what you can find out about how it works. https://en.wikipedia.org/wiki/Determination_of_the_day_of_the_week#Sakamoto.27s_algorithm

c. The display will sometimes include a "*" before the time; investigate in the code to find the conditions for this character being displayed.

3. Investigate the addressing of the I2C LCD interface, change the device address using the solder links and modify the code accordingly.
4. Look at the use of the variable SerialDebug, change it from FALSE to TRUE and then connect your PC to the USB to Serial interface and monitor the traffic. A good PC application to monitor serial port communication is Tera Term. Add some extra conditional serial comms elsewhere in the code to send additional information to the PC terminal screen.
5. Parsing Strings and manipulation of time and date are useful concepts; investigate the experimental routines in the project and find other ways to reuse the routines in the code for your own applications.

10

STM32CubeIDE Project
A GPS disciplined frequency counter

Shopping List

This section of the book uses the following components (Please refer to the hardware list file on the Books extra page for this book to obtain up-to-date URLs for where to buy these items):

- Nucleo-F401 Module

Introduction

We are going to use a STMicroelectronics Nucleo board for this project; the board in question contains a STM32F401RE microcontroller and an integrated STLINK programming and debug interface.

The concept of using a GPS to discipline an oscillator or frequency counter is not new. Many individuals have spent significant time developing complex algorithms to maximise the accuracy associated with this kind of configuration and the details can be complex. Here we are going to learn the basics of the concept and develop an experimental frequency counter using GPS disciplining.

The GPS module that we utilised in project 1, the GPS synchronised clock, contains a 1PPS signal output that we have not yet utilised. This signal provides an accurate 1 pulse-per-second

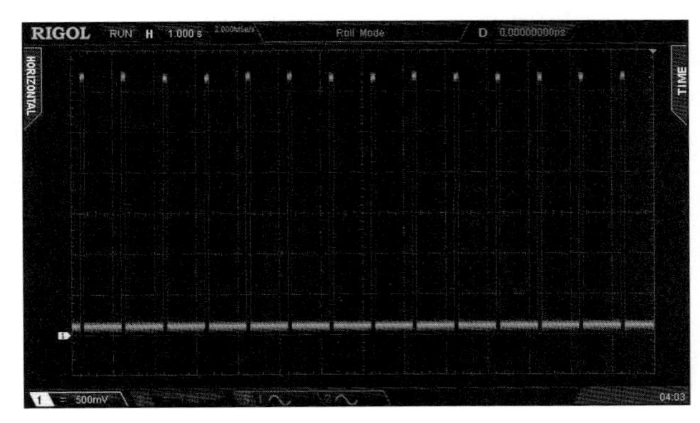

Figure 10.1 - 1pps GPS output

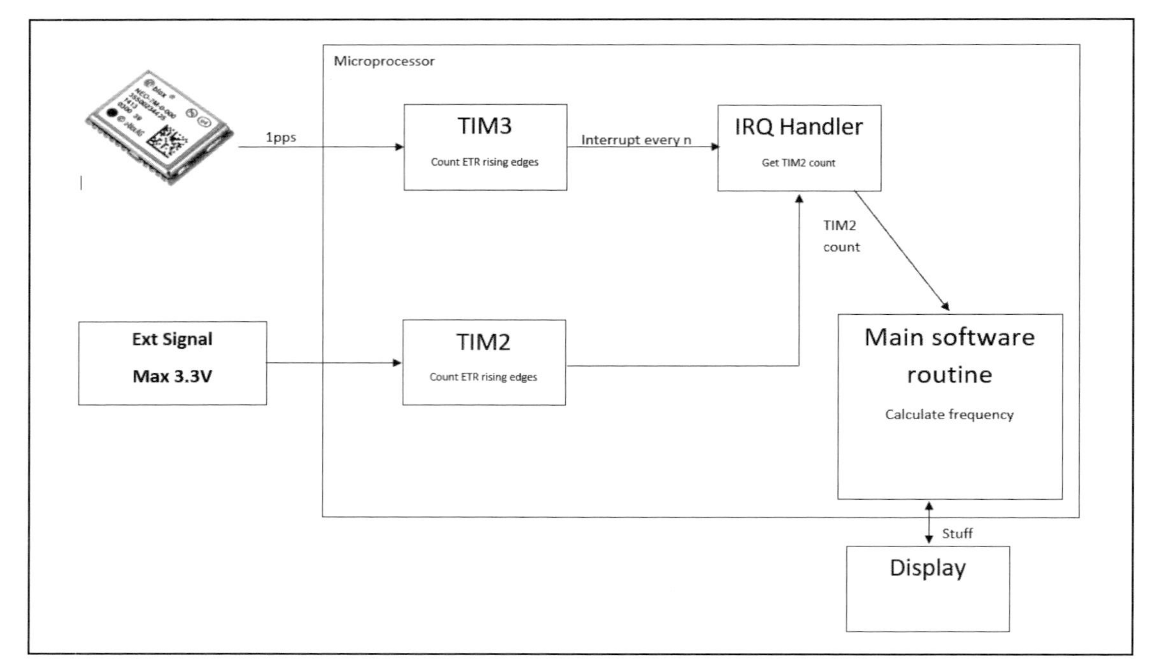

Figure 10.2 - GPS Frequency Counter Overview

timing signal that we will use in this project. The output from the PPS signal is shown in **Figure 10.1**.

The basic concept for this project is illustrated in **Figure 10.2**.

For the first time we will use timers within the STM32 microcontroller, one timer will count the 1pps signals from the GPS module and generate a software interrupt at a configurable count, the second timer will count an external signal of unknown frequency.

Each time the software is interrupted by the 1PPS counter, we need to fetch the count from the second timer counting the external signal. As we know the time interval between the fetches of this counter value, we can calculate the frequency of the external signal. If desirable, we can take multiple readings of the frequency and average these to calculate a result.

As implied earlier, this is an over simplistic method of frequency counting as the GPS data may be subject to jitter and far more complex algorithms are needed for extreme accuracy. However, we are presenting a concept here, and we will now progress with my favourite method of software development; write a bit, test a bit, repeat.

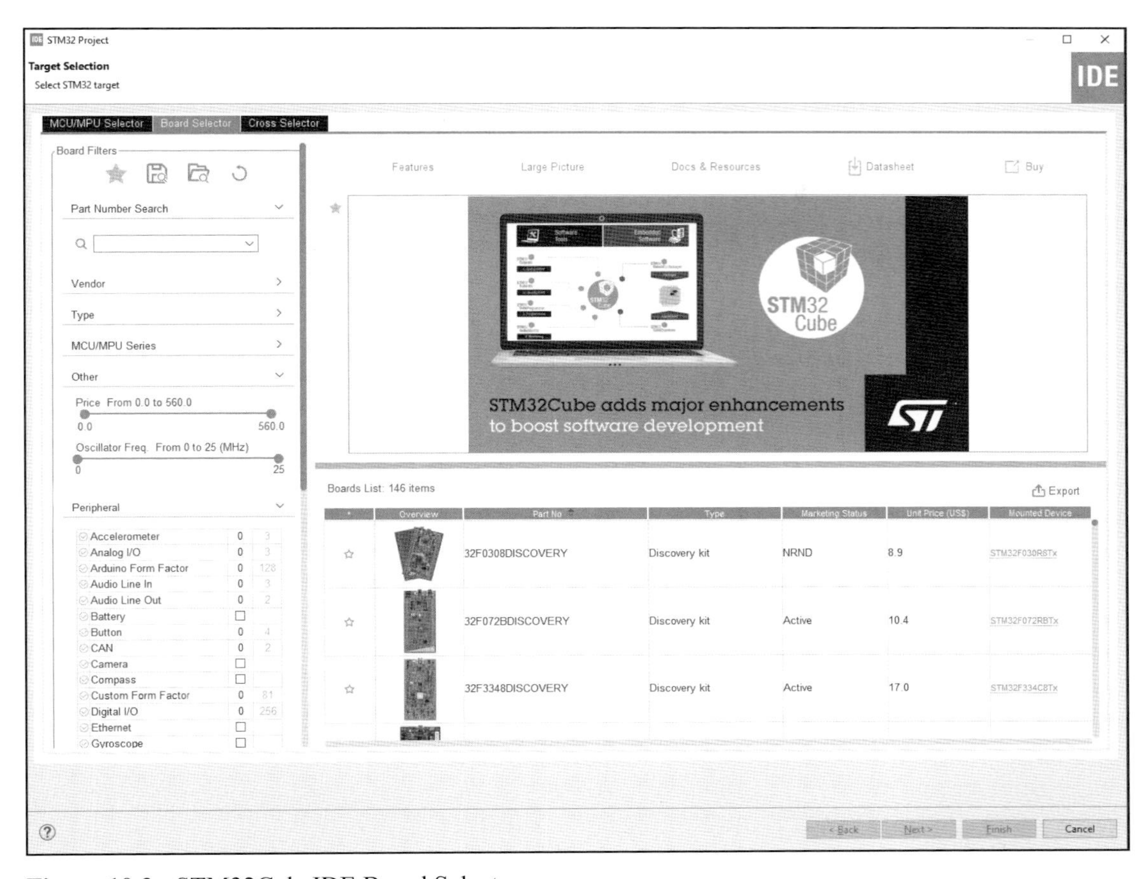

Figure 10.3 - STM32CubeIDE Board Selector

STM32CubeIDE Configuration

The STM32 Cube IDE has in-built configurations for all the available Nucleo range of boards, so our setup differs from the previous project as a lot of the work will be done for us automatically by the IDE.

Start a new project in the IDE by using the File Menu and New, then select STM32 Project. This will bring up the Target Selection window we used before. This time we will select the "Board Selector" tab as shown in **Figure 10.3**.

Within the search box type "F401" and the selections available will be reduced and the Nucleo board we are using should be listed in the Board List as shown in **Figure 10.4**.

Click on the NUCLEO-F401RE board in the list and click next.

You will be prompted to give the project a name, as previously we will prefix the

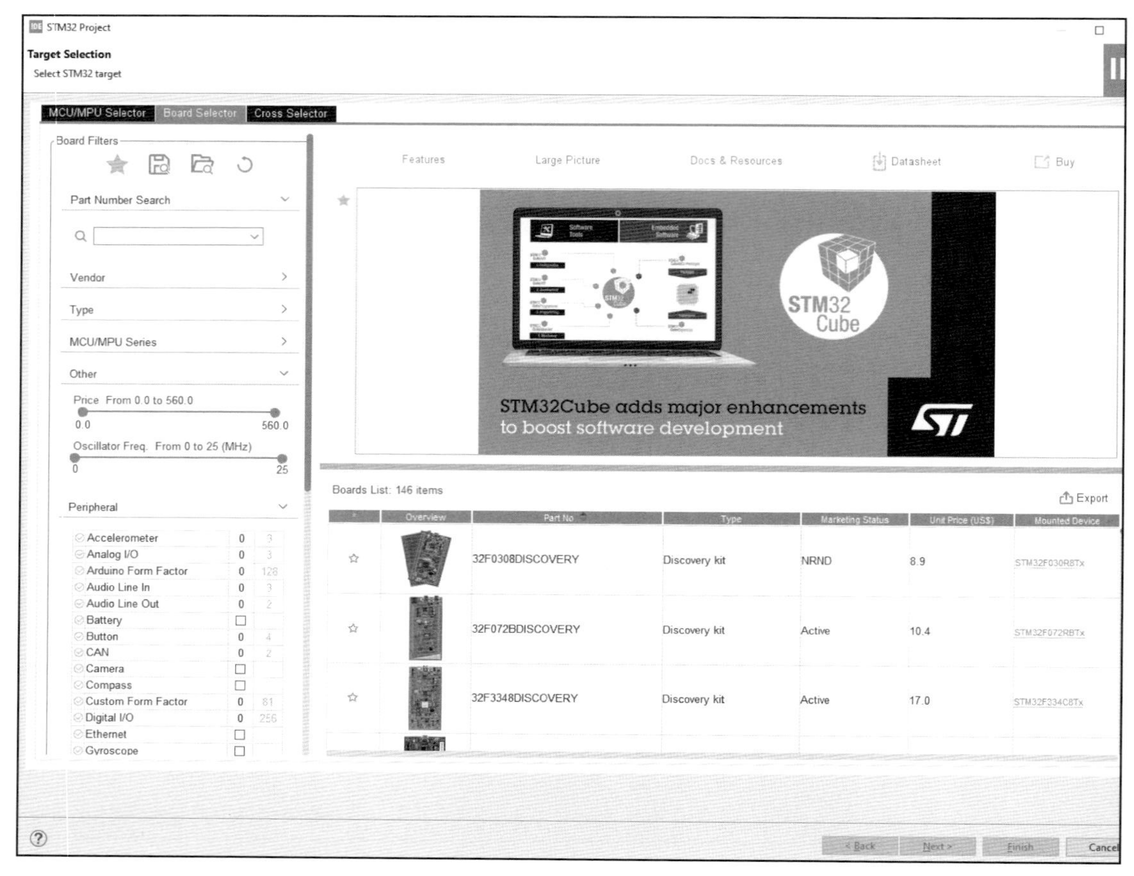

Figure 10.4 - STM32F401 Board List

project with the processor type, so here we will call the project STM32F401RE-GPS-Freq-Count. Once you have entered the project name, leave all the other options at their defaults and click finish.

You will be prompted to "Initialise all peripherals to their default mode?" - answer Yes. This will configure the MCU to work with our Nucleo board.

A word about Timers

All of the STM32 MCUs contain a number of timers. As far as the MCU is concerned these are simply microcontroller peripherals like anything else. The basic timer construction is shown in **Figure 10.5**.

Timers can be used for a wide range of applications including:

- Time base generation
- Measuring/counting external signals

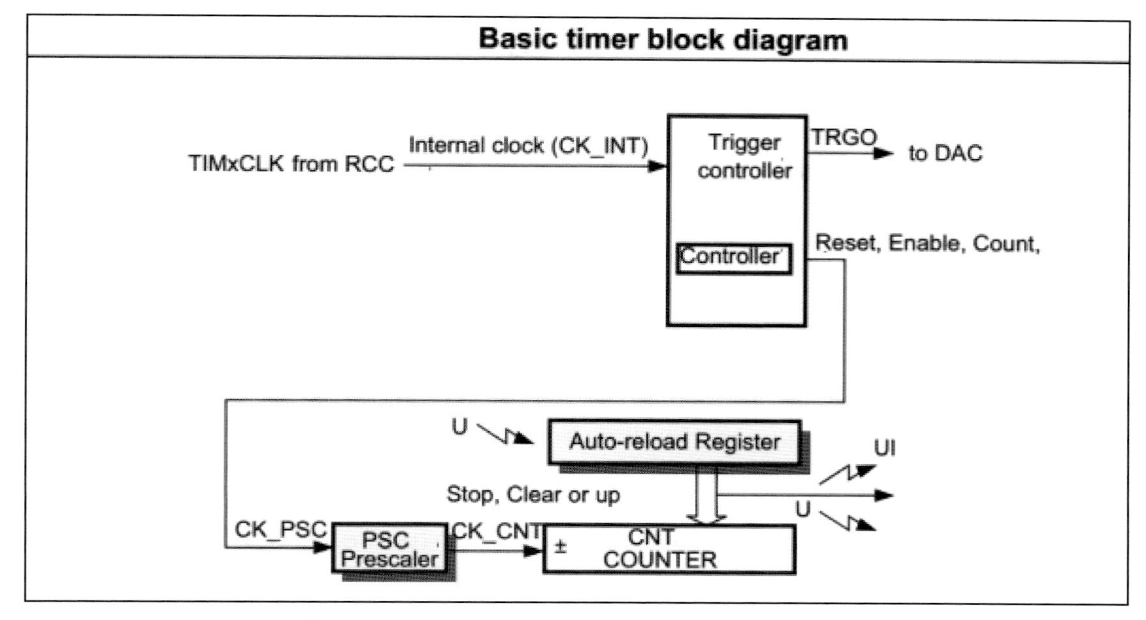

Figure 10.5 - STM32 Basic Timer Block Diagram

- Producing waveforms
- Measuring pulse widths
- Generating Pulse Width Modulation (PWM) signals
- Triggering external devices

There are three fundamental types of timer in the STM32 MCU range:

- Basic Timers (most STM32 MCUs contain one or more of these)
- General Purpose Timers (most STM32 MCUs contain at least one of these)
- Advanced Timers (some STM32 MCUs contain one or more of these)

Timers can be 16-bit (able to count from 0 to 0xFFFF hex) or 32-bit (able to count from 0 to 0xFFFFFFFF hex).

Our selected MCU, the STM32F401RE, contains the following:

- TIM1 (Timer 1) – Advanced Timer
- TIM2 & TIM5 – 32-bit General-Purpose Timers
- TIM3, TIM4, TIM9/10/11 – 16-bit General-Purpose Timers

We are going to utilise TIM3, a 16-bit timer, to count our GPS 1pps signal and TIM2, a 32-bit timer, to count our external frequency source.

Step 1 – External 1pps Counting

Project Configuration

We need to configure TIM3 to count our external GPS 1pps signal. As explained above, we need to route the PPS signal into the TIM3 timer. On the IDEs graphical pinout view of our MCU, find pin PD2 and left click to see the list of possible configuration options; select TIM3 ETR.

Within the Timers section of the IDE, select TIM3 and set the clock source to be ETR2.

In the configuration section under the NVIC tab, select the tick box to enable the TIM3 Global Interrupt.

Under parameter settings set the counter period to be 5.

Save the file and generate the code.

This will configure the TIM3 to count the external 1PPS signal and generate an interrupt every count of 5.

We now need to provide a callback routine to handle the interrupt and perform the tasks required. Following my concept of write a bit, test a bit, repeat, here we will simply toggle the on board LED every time the interrupt is generated as our step 1. The callback routine is already generated by the IDE for us and is called HAL_TIM_PeriodElapsedCallback and is declared as 'weak' which means we can re-define, or overwrite if you like, the routine in our code.

Within the User Code 4 section of the main.c file we can now add our own version of the callback routine. The onboard LED of our Nucleo Board is connected to GPIO PortA pin 5.

```
void HAL_TIM_PeriodElapsedCallback(TIM_HandleTypeDef
*htim)
{
  // Toggle the LED
  HAL_GPIO_TogglePin(GPIOA, GPIO_PIN_5);
}
```

Our final piece of code needed to execute step one is to start the TIM3 timer. In the user code 2 section we need to add:

```
//TIM3 is interrupting on the external PPS signal
HAL_TIM_Base_Start_IT(&htim3);
```

This Step 1 code is included in the Books extra page and can be imported as a complete project.

Project Build

Using the pinout diagram of the Nucleo 64 board contained in the Books extra page in directory 'Nucleo 64 pinout', connect the 1pps signal from the GPS to the PD2 pin (pin 4 of CN7) and use the 3V3 (pin 4 of CN6) and GND (pin 6 of CN6) to power the GPS.

Click on the Play Icon to compile, download and execute the code on the target device.

You should see our on-board LED toggling for every 6 pulses of our 1pps signal from the GPS.

Things to Try

1. Alter the number of counts the timer completes before the LED is toggled.
2. Explore and understand why when we have set the counter period to be 5, the LED is toggling every 6 pulses of the 1 PPS.
3. The toggle of the LED is occurring on the rising edge of the 1pps signal. Explore ways to change this to be the falling edge.

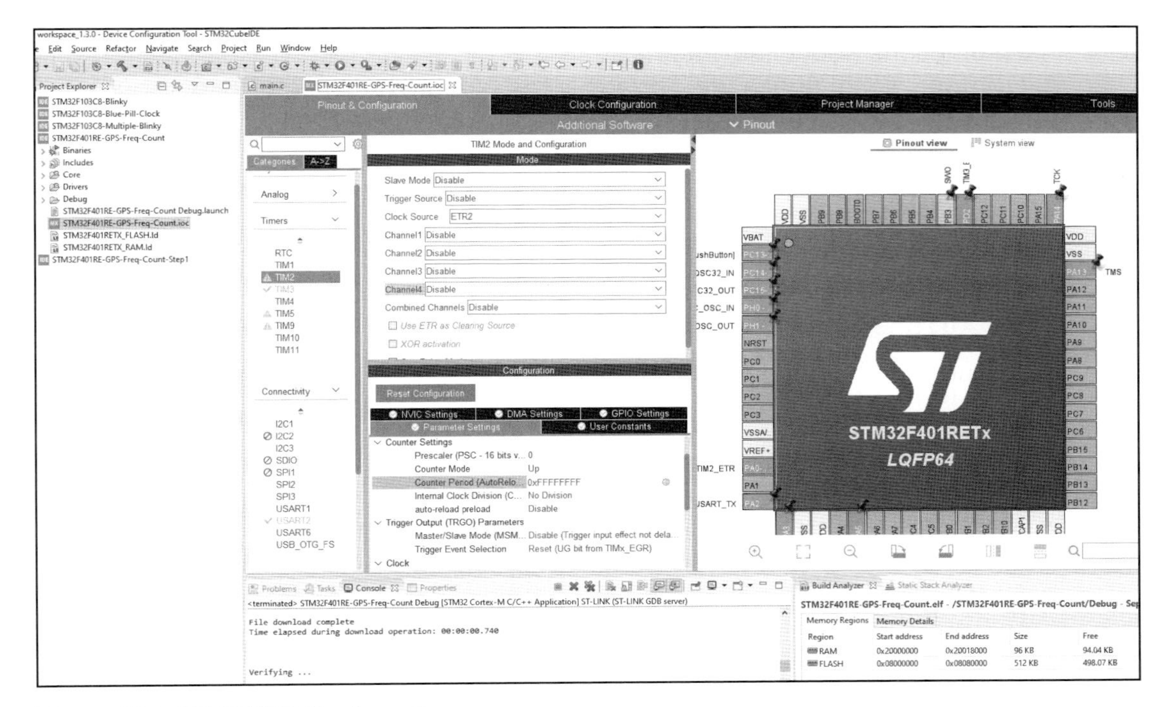

Figure 10.6- TIM2 ETR Configuration

Step 2 – Counting the External Signal

We will now select TIM2 in the IDE, configure the clock source to be ETR2 and set the Counter Period to be the maximum for a 32 bit counter, 0xFFFFFFFF Hex. This is shown in **Figure 10.6**. This will configure Timer2 to count the signal on pin PA0 of the MCU with the counter wrapping round when it gets to the maximum value a 32-bit register can store.

Remember that this is a 3V3 device so any signal applied directly to the GPIO pins must comply with the voltage restrictions to avoid damaging the device.

In a real-world application we would include external signal conditioning hardware prior to connecting to the MCU pin.

If you haven't done so already, change the counter period of TIM3 to be 4 so we get our interrupt every 5 1pps pulses, and once you have also configured TIM2 as above, save the project and generate the code.

TIM2 is now configured to count an external signal, it is a 32-bit timer so will count from 0x0 Hex to 0xFFFFFFFF hex and then wrap round back to 0x0.

We need to modify our TIM3 interrupt callback routine to read the counter value from TIM2 and make it available in our main code.

Any variables that we declare in our main code that can be updated by interrupt routines must be declared as 'volatile'. This prevents the compiler from doing clever things and performing code optimisation. Optimisation would be a bad idea as this variable could be updated at any time, so the code must read the value from its memory location every time it's needed, and not make any assumptions about its contents.

In our global space we declare some global variables for use in our main routine:

```
/* USER CODE BEGIN PV */

volatile uint32_t CurrentExternalCount = 0;
uint32_t OldExternalCount = 0;
uint32_t TotalCount = 0;
uint32_t ExternalFrequency = 0;
uint32_t SampleTime = 0;

/* USER CODE END PV */
```

Note the use of the keyword volatile for the variable CurrentExternalCount; this value will be updated by our interrupt callback routine.

Within User Code section 2 we have already added a line to start the 1pps counting timer, we now need to also add a line of code to start the counter for the external signal:

```
/* USER CODE BEGIN 2 */

HAL_TIM_Base_Start_IT(&htim3);// TIM3 is the 1pps
                              // counter
HAL_TIM_Base_Start(&htim2);   // TIM2 is counting our
                              // external signal

  /* USER CODE END 2 */
```

Within our existing callback routine we now need to add the code that will copy the counter value from TIM2 into the volatile variable we declared above:

```
/* USER CODE BEGIN 4 */

void HAL_TIM_PeriodElapsedCallback(TIM_HandleTypeDef
*htim)
{
        HAL_GPIO_TogglePin(GPIOA, GPIO_PIN_5);
        CurrentExternalCount = __HAL_TIM_GET_
        COUNTER(&htim2);
}

/* USER CODE END 4 */
```

So now, every time we get our 1pps counter interrupt, the variable CurrentExternalCount will be updated with the value from the timer that is counting our external signal.

Within our main loop we now need to decide if this value has been updated and if so we should then proceed to calculate the frequency of the external signal.

My code to calculate the frequency looks like this:

```
  /* USER CODE BEGIN WHILE */
  while (1)
  {
        // CurrentExternalCount is updated by the
        // interrupt routines
        // so if its changed outside this loop we have
        // counting
        if (CurrentExternalCount != OldExternalCount)
        {
              // calculate the delta count between last
              // time and this
              if (CurrentExternalCount > OldExternalCount)
```

```
                {
                    TotalCount = CurrentExternalCount -
                    OldExternalCount;
                }
                else
                {
                    // count has wrapped round in the
                    // timer
                    TotalCount = 0xFFFFFFFF -
                    OldExternalCount +
                    CurrentExternalCount;
                }

                    // set the old count value for next
                    // time
                    OldExternalCount = CurrentExternalCount;
                    // Total Count now contains the
                    // number of counts between the
                    // last two 1pps interrupts
                    // now we need to calculate the
                    // frequency

                SampleTime = htim3.Init.Period + 1;
                ExternalFrequency = TotalCount /SampleTime;
            }

        /* USER CODE END WHILE */
```

The only remaining piece of the puzzle is to communicate this calculated frequency. We will utilise the COM Port provided by the Nucleo board to do this. The IDE will have already declared and initialised UART2 for this purpose, we now need a simple routine to utilise the COM port to communicate with the PC.

The communication routine looks like this:

```
void PrintSerial(char *format,...)
{

char str[80];

/*Extract the the argument list using VA apis */
      va_list args;
      va_start(args, format);
      vsprintf(str, format,args);
```

```
        HAL_UART_Transmit(&huart2,(uint8_t  *)str,
       strlen(str),HAL_MAX_DELAY);
        va_end(args);
}
```

Note this is an extremely useful routine and can be re-used in your projects going forward whenever you need serial debug communications with your PC.

The routine above is added to the USER CODE 4 section of our main.c and the prototype of the function declared thus:

```
/* USER CODE BEGIN PFP */

void PrintSerial(char *format,...);

/* USER CODE END PFP */
```

The routine utilises several standard C functions which need to be made available to main.c through include statements:

```
/* USER CODE BEGIN Includes */

#include <string.h>
#include <stdio.h>
#include <stdarg.h>

/* USER CODE END Includes */
```

Every time we calculate the frequency, we will now send the result to the COM port by using these additional lines of code:

```
sprintf(UsrString,"Frequency is %ld\r\n",ExternalFrequency);
PrintSerial(UsrString);
```

The variable UsrString is added to our global space:

```
char UsrString[100] = {0};
```

Sprint is a very versatile string manipulation function included in the C language, we create a single string using this function and then use our PrintSerial routine to send the string in its entirety to the serial port.

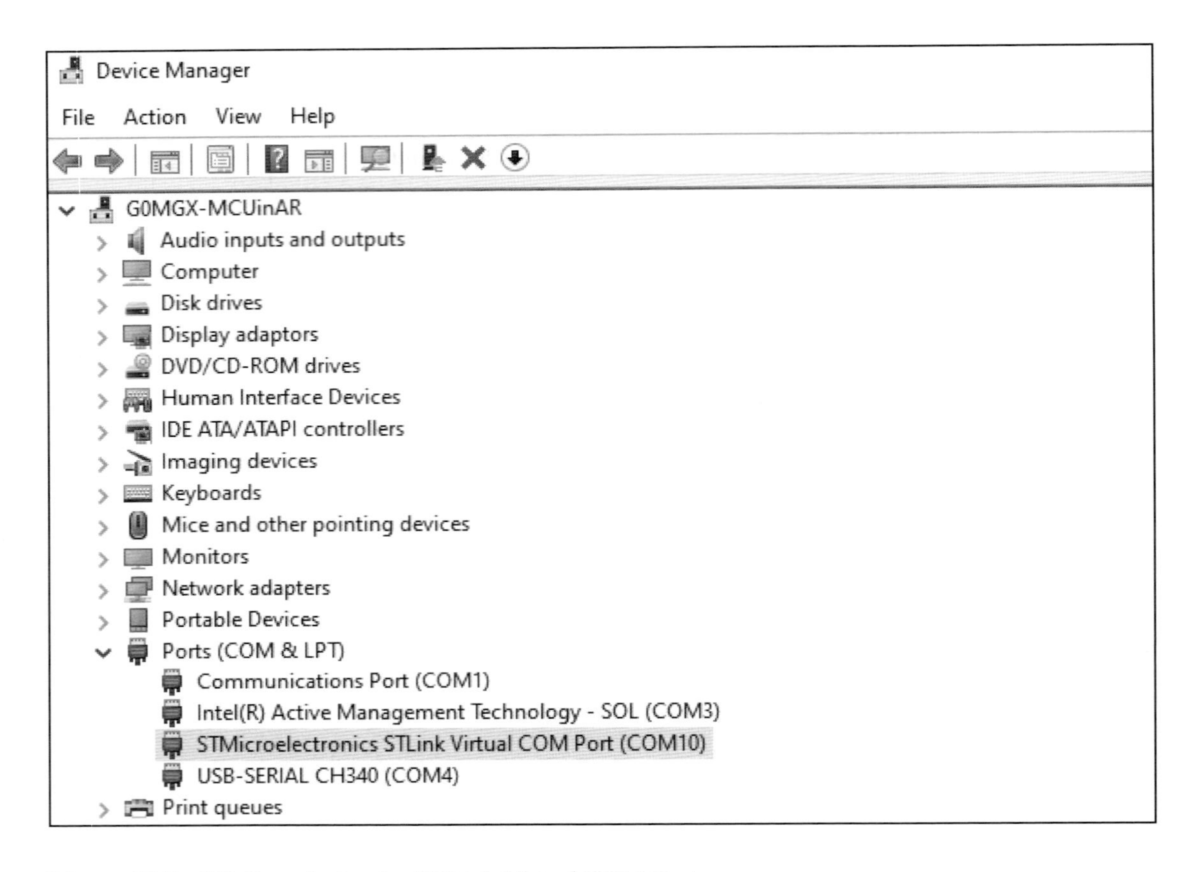

Figure 10.7 - STMicroelectronics STLink Virtual COM Port

Project Build

Referencing the pinout diagram of the Nucleo 64 board contained in the Books extra page in directory 'Nucleo 64 pinout', connect the 1pps signal from the GPS to the PD2 pin (pin 4 of CN7) and use the 3V3 (pin 4 of CN6) and GND (pin 6 of CN6) to power the GPS.

Click on the Play Icon to compile, download and execute the code on the target device.

You should see our on-board LED toggling for every 5 pulses of our 1pps signal from the GPS.

Open a terminal program, (Tera Term is recommended) and configure it to use the STMicroelectronics STLink Virtual COM Port, Windows Device Manager shows an example in **Figure 10.7**.

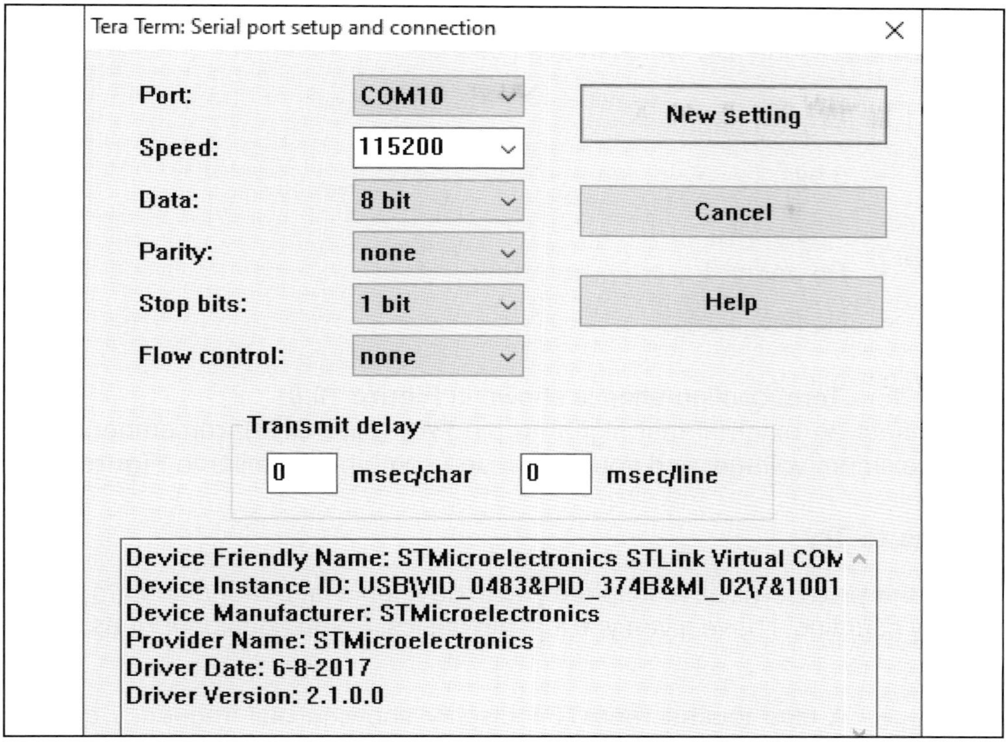

Tera Term: Serial port setup and connection

Port: COM10
Speed: 115200
Data: 8 bit
Parity: none
Stop bits: 1 bit
Flow control: none

New setting
Cancel
Help

Transmit delay
0 msec/char 0 msec/line

Device Friendly Name: STMicroelectronics STLink Virtual COM
Device Instance ID: USB\VID_0483&PID_374B&MI_02\7&1001
Device Manufacturer: STMicroelectronics
Provider Name: STMicroelectronics
Driver Date: 6-8-2017
Driver Version: 2.1.0.0

Figure 10.8 - Tera Term Configuration

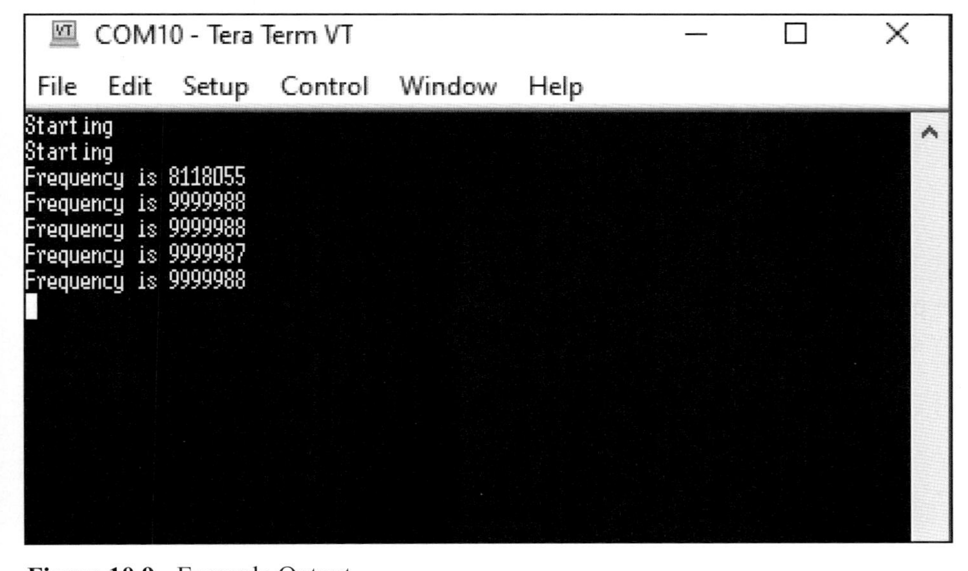

Figure 10.9 - Example Output

Make sure you configure Tera Term to the same as the default COM settings in the STM32 Cube IDE which in this case are:

- 115200 baud
- 8 bits
- 1 stop bit
- No parity
- No flow control

The Tera Term configuration is shown in **Figure 10.8**.

Connect an external test signal to pin PA0 (pin1 CN8), remembering the 3V3 maximum voltage, and you should see something similar to **Figure 10.9**.

Things to Try

I hope this project will have inspired you to develop a far better algorithm for frequency counting, I have given you just the basics. Some additional ideas follow:

- A nice touchscreen TFT display showing our frequency counter output would be a good addition
- Have a look at the pre-scaler options for the Timer that's counting our external signal; there are many possible configurations using a pre-scaler that will enable very high external frequency signals to be counted.
- You could use the touch screen TFT to allow the user to select the pre-scaler applied to the signal.
- You could also provide the user the ability to configure the "gate" time, which in our code is currently fixed at 5 seconds.

The possibilities of expanding this basic project into an accurate shack frequency counter are almost endless.

ESP8266

11

Overview

ESP8266 is a low-cost, low-power microcontroller which contain integrated Wi-Fi making it ideal for the Internet of Things (IoT) market. The ESP8266 series employs a Tensilica Xtensa Diamond Standard 106Micro running at 80 MHz and has a built-in TR switch, balun, LNA, power amplifier and matching network.

ESP8266 was created and developed by Espressif Systems, a Shanghai-based Chinese company, and is manufactured by TSMC. It is a predecessor to the highly successful ESP32 microcontroller, which was released in 2016.

Development Boards

A large number of boards are available in the market-place based on both the older ESP8266 MCU and the more recent ESP32 range of MCUs.

We are going to use a readily available ESP8266 board.

This board contains the DOITING ESP-12F MCU and is shown in **Figure 11.1**.

The datasheet for the MCU is contained in the book file repository.

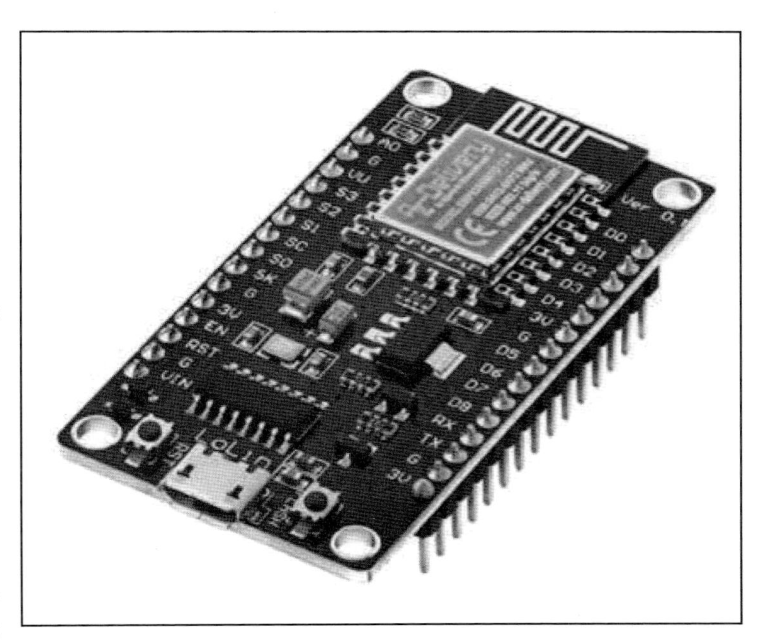

Figure 11.1 - MCU ESP-12F Development Board

Preferences

Settings Network

Sketchbook location:

C:\Users\markg\OneDrive\Documents\Arduino Browse

Editor language: System Default (requires restart of Arduino)

Editor font size: 12

Interface scale: Automatic 100 % (requires restart of Arduino)

Theme: Default theme (requires restart of Arduino)

Show verbose output during: compilation upload

Compiler warnings: None

Display line numbers

Enable Code Folding

Verify code after upload

Use external editor

Check for updates on startup

Save when verifying or uploading

Use accessibility features

Additional Boards Manager URLs: :kage_STM32duino_index.json, https://arduino.esp8266.com/stable/package_esp8266com_index.json

More preferences can be edited directly in the file

C:\Users\markg\AppData\Local\Arduino15\preferences.txt

(edit only when Arduino is not running)

OK Cancel

Figure 11.2 - Arduino IDE board manager URLs

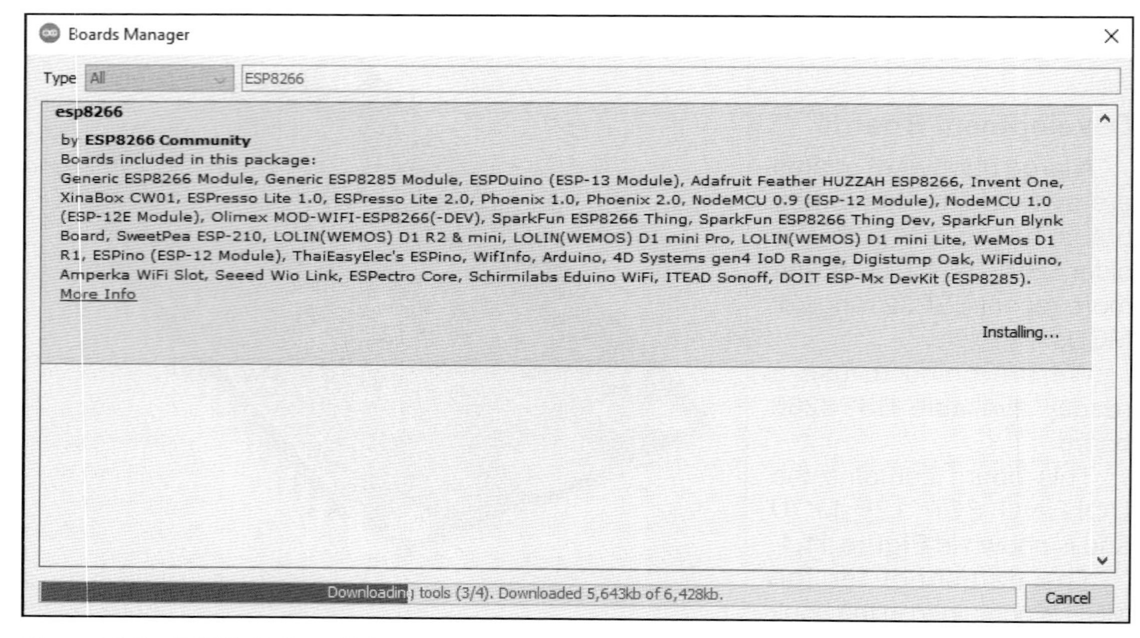

Figure 11.3 – ESP8266 Board Manager

Installing the ESP8266 Arduino Core

We will go back to using the Arduino IDE for our ESP8266 project and install the board support package. In a similar process to that which we followed when adding the STM32 processor boards to the IDE, we will add the following to the board manager URLs section. If you already have entries in this configuration box, additional entries need to be separated by commas.

https://arduino.esp8266.com/stable/package_esp8266com_index.json

Start the Arduino IDE and select File > Preferences then add the URL above into the "Additional Board Manager URLs" section. An example is shown in **Figure 11.2**.

Close the Preferences box and then go to Tools > Board > Boards Manager and search for ESP8266 as shown in **Figure 11.3**.

Click the Install button to add these boards to the IDE.

Once Installed you should be able to select the target board 'Espino ESP-12'.

Once you connect your ESP8266 board to your PC you should see a new COM port within the Microsoft device manager as shown in **Figure 11.4**.

Select the COM Port number for the ESP8266 within the IDE in Tools > Port.

As we did with our STM32 Blue Pill board, we should now be able to compile and execute a test program. We will use the provided example 'Wi-Fi scan' which is found in File > Examples > ESP8266Wi-Fi -> Wi-Fi Scan.

Use the upload button in the IDE to compile and upload the program to the board.

Once uploaded open the Serial Monitor in the IDE from Tools > Serial Monitor and change the baud rate to 115200, press the reset button on the ESP8266 board and you should see an output similar to that shown in **Figure 11.5**.

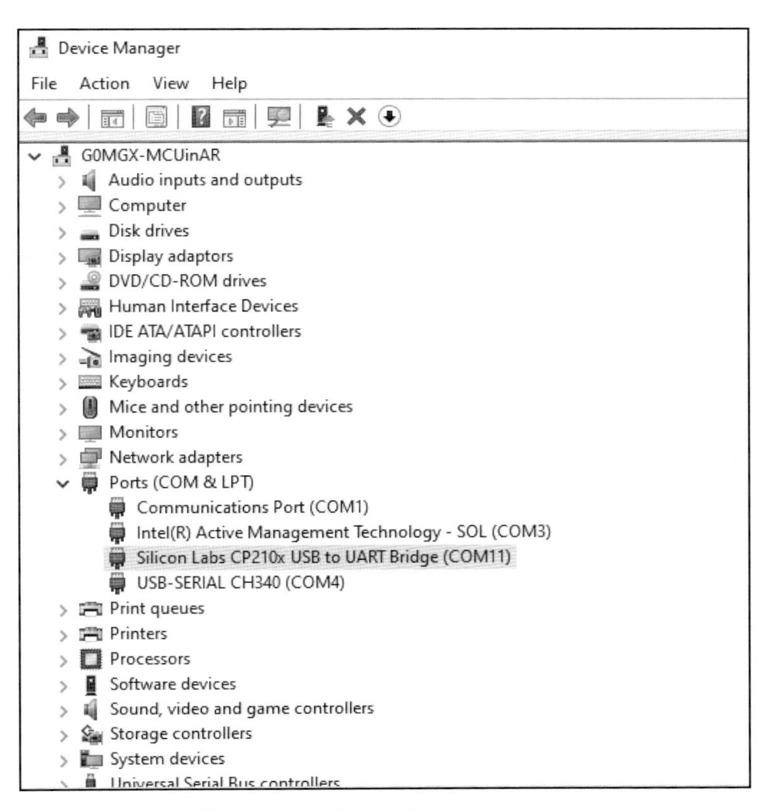

Figure 11.4 – ESP8266 COM Port Entry

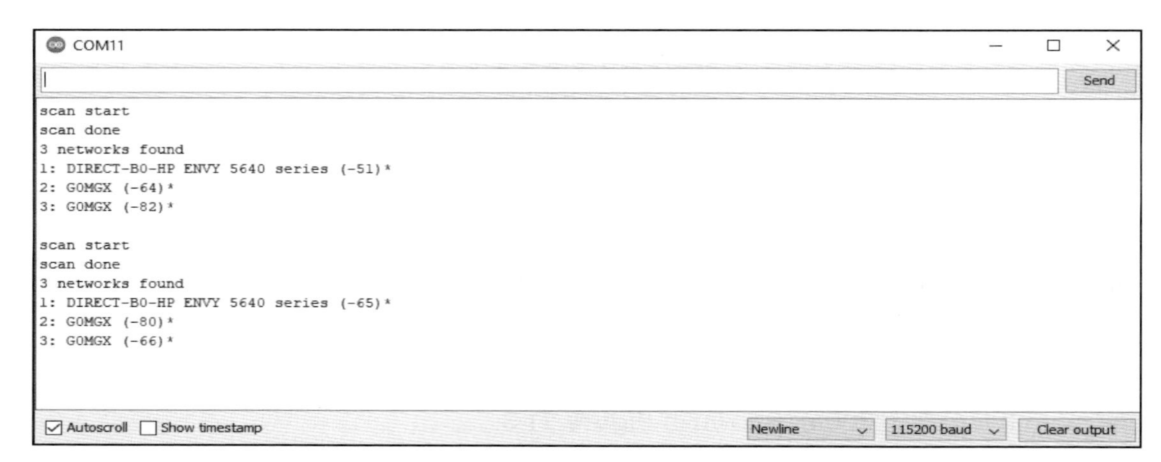

Figure 11.5 – ESP8266 Wi-Fi Scan Output

We can conclude from the above output that our board is installed and working correctly. We can now move on to our project which is an Internet connected Solar Monitor for the shack.

12

ESP8266 Project
A Shack Solar Conditions Monitor

Shopping List

This section of the book uses the following components (Please refer to the hardware list file on the RSGB Book File Repository for this book to obtain up-to-date URLs for where to buy these items):

- ESP8266 Module
- BME280 Sensor
- ILI9341 TFT Display

Introduction

We are going to build a shack based solar conditions display. This project will download solar data from the internet and display the information on a TFT screen. The BME280 sensor included in the project will allow us to also determine temperature (where the sensor is located) and the air pressure. Three separate screens of information will be presented cyclically, the solar data, the predicted band conditions and the temperature.

This project utilises some of the concepts presented here:
http://www.variableindustries.com/web-to-lcd-2/

Project Configuration

Wire up your ESP8266 module with the TFT and the BME280 sensor as shown in **Figure 12.1**.

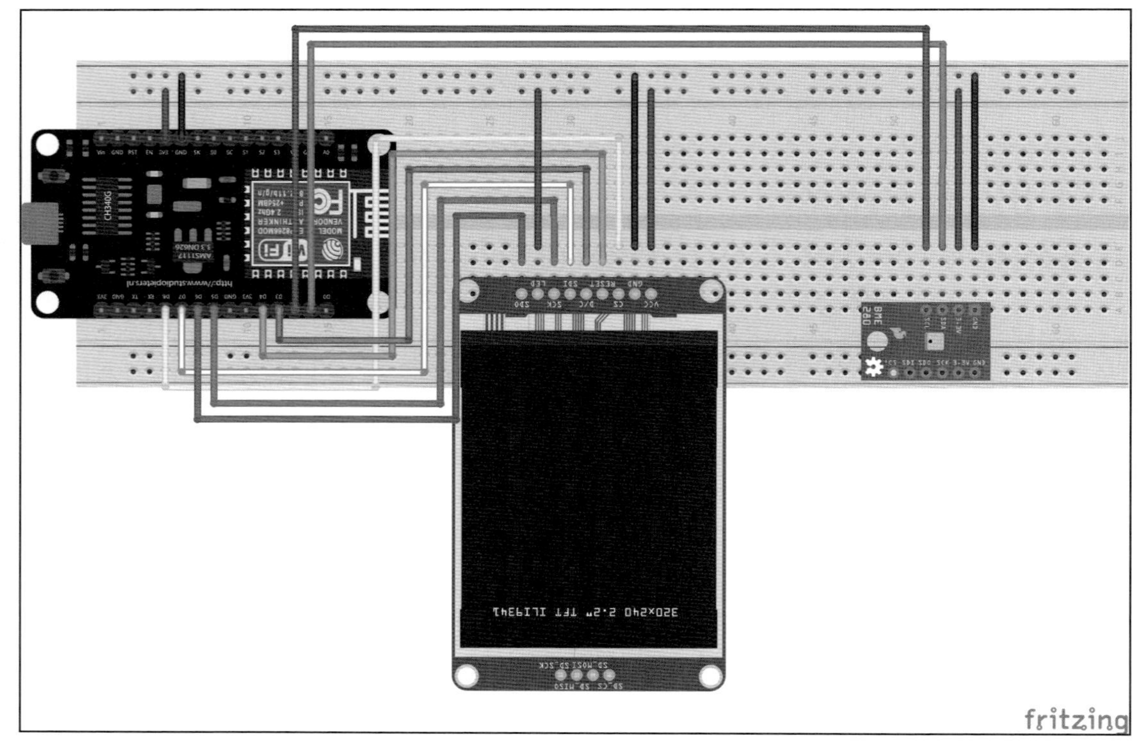

Figure 12.1 - ESP8266 Project Configuration

Project Build

Once you have opened the source code we can go ahead and build our project.

You should be familiar with the Arduino IDE by now and will need to use Library Manager to install the AdaFruit_BME280 library used by the code. Once you have your project compiled correctly, use the CTRL-U shortcut to upload the project to the ESP8266 target device. Once upload is complete, you need to connect to the ESP8266 from an internet browser to configure the Wi-Fi connection.

The ESP8266 will create a Wi-Fi 'hot-spot' that you can connect to using a mobile phone or other device. The ESP8266 is configured to have an IP address of 192.168.0.1 by these lines of code:

```
IPAddress apIP(192, 168, 0, 1);
IPAddress netMsk(255, 255, 255, 0);
```

And have a network name of "ESP LCD" and password of "banana" by these lines of code:

```
const char *ESPssid     = "ESP LCD";
const char *ESPpassword = "banana";
```

Once you have connected to the ESP8266 network, follow the on-screen instructions to enter your Wi-Fi network credentials. The software will then store your Wi-Fi network credentials in non-volatile memory and the ESP8266 will in future, automatically connect to your Wi-Fi network.

Feel free to alter the code to bypass this step and hard code the Wi-Fi credentials into the software if you wish.

Once connected, you should see the TFT alternate the display between Solar Data, Band Predictions, and temperature screens. Note: debug information is available through the Serial Monitor running at 115200 baud.

Welcome to the Internet of Things!

Microchip ARM

13

Overview

Microchip/Atmel ARM is a family of MCUs produced by Microchip (previously from Atmel up until 2016) which are based on various 32-bit ARM cores and Atmel/Microchip designed peripherals and tool support.

As we learned previously, ARM produce and licence the core which is then taken by other manufacturers to produce MCUs for the market.

This range includes the SAM C, SAM D, SAM L, SAM 3, SAM 4 and SAM x70 MCUs all of which are 32-bit devices. There are also devices designed for specific market applications such as the SAM4CP which is used exclusively in smart meters.

Microchip ARM MCUs range from the SAM D10 series with as few as 14 pins, to the 144-pin SAM S70 and SAM E70 products.

Development Boards

There are many development boards available for the Microchip ARM range of MCUs, for our experiments we will use the Arduino Zero which contains the ATSAMD21G18 MCU.

The Arduino Zero is available as a genuine Arduino product and as a number of clone boards which operate in exactly the same way.

Figure 13.1 shows some of the authors development boards, a copy of the Arduino Zero is on the right, the centre is a manufacturer board, and the left shows a "SAMD21 Mini" which is an inexpensive development board available from many outlets.

For our experiments we will use the "SAMD21 MINI" board with the Arduino IDE. A closer look at the board is shown in **Figure 13.2**.

Figure 13.1 - Atmel SAMD Boards

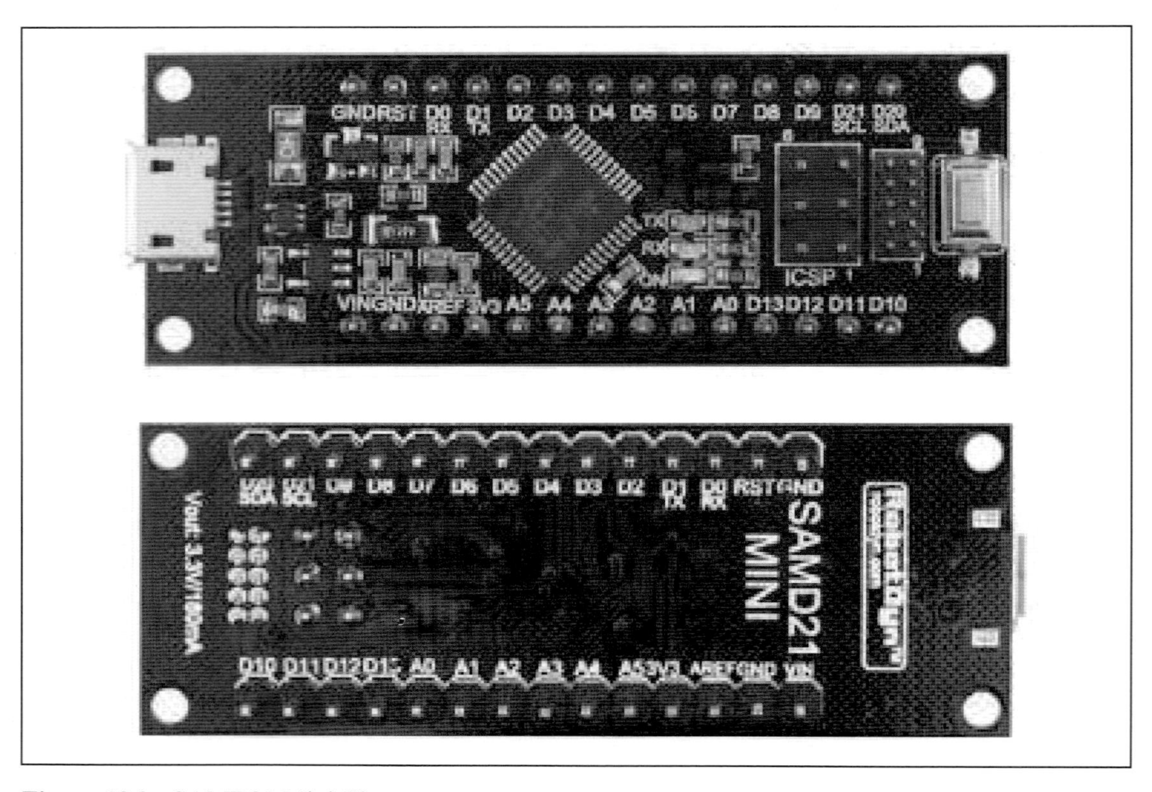

Figure 13.2 - SAMD21 Mini Closeup

Arduino Support for SAMD

Open up your Arduino IDE and navigate to Tools->Board Manager. In the search box type in "SAMD" as shown in **Figure 13.3**.

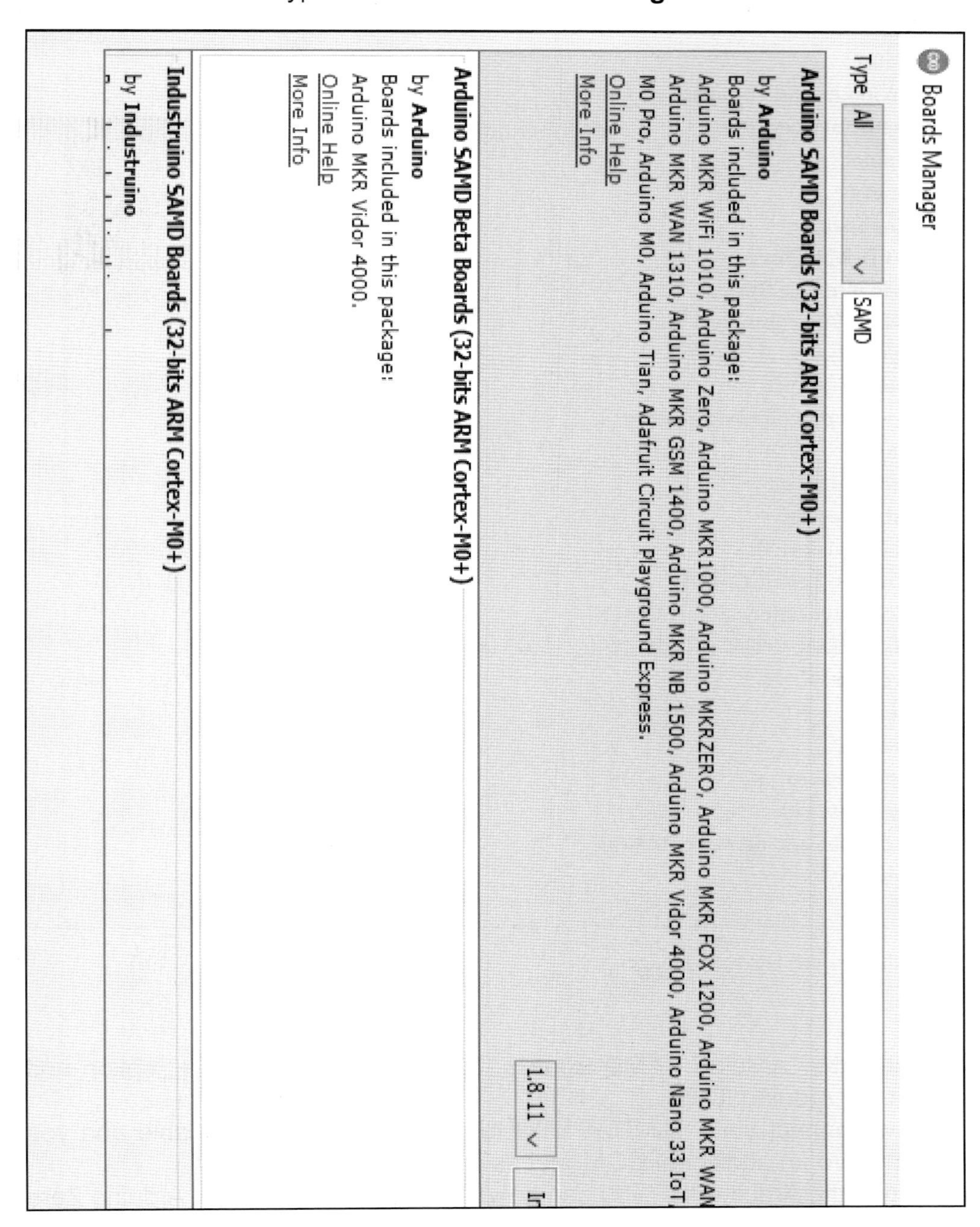

Figure 13.3 - Arduino Support for SAMD

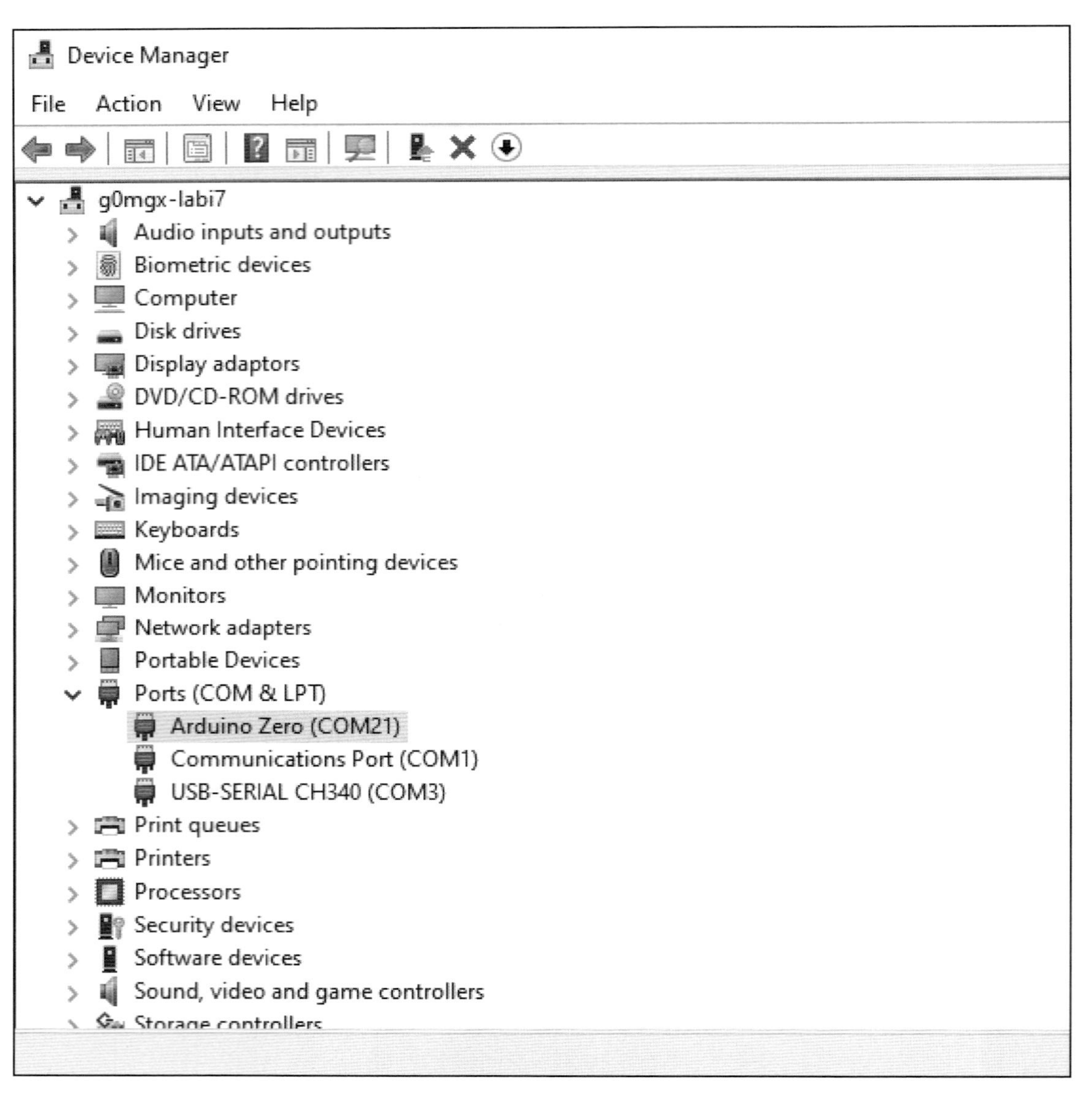

Figure 13.4 - Arduino Zero Device Manager

Install the "Arduino SAMD Boards (32-bits ARM Cortex-M0+" option and restart the Arduino IDE.

Now connect your SAMD21 MINI to your PC with a USB cable and you should have an additional COM port as shown in **Figure 13.4**.

Within the Arduino IDE you can now select the board type of Arduino Zero (Native Port) as shown in **Figure 13.5**.

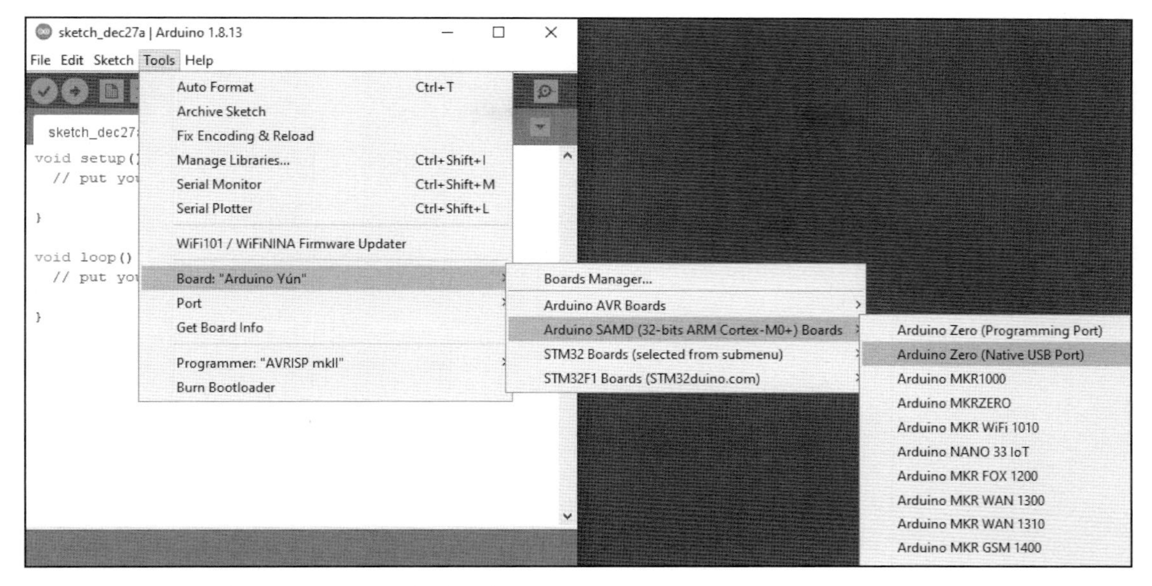

Figure 13.5 - Arduino SAMD Board Selection

You also need to select the port number of your Arduino Zero as displayed in Device Manager in the Tools->Port selection.

Blinky Once More

There is no on-board LED on the SAMD21 MINI, but we can test our board by uploading a simple sketch to toggle pin D7 every second.

```
void setup() {
  pinMode(7, OUTPUT);
}

// the loop function runs over and over again forever
void loop() {
  digitalWrite(7, HIGH);
  delay(1000);
  digitalWrite(7, LOW);
  delay(1000);
}
```

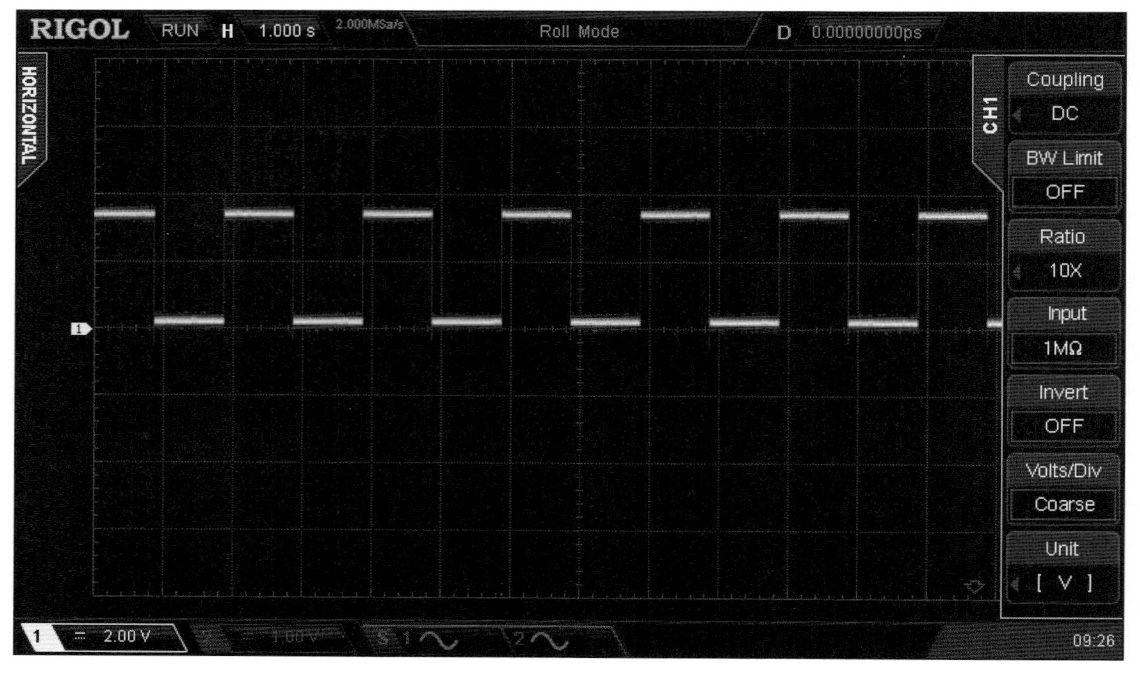

Figure 13.6 - Pin D7 Output

You can confirm the correct operation of your board either by connecting an LED and series resistor to pin D7 or using an oscilloscope as shown in **Figure 13.6**.

Congratulations – you have your first 32-bit MCU application running.

14

SAMD21 Project
Programming the ADF5355

Modern electronics often amaze me; none more so than the constant evolution of Direct Digital Synthesis (DDS) chips, especially those produced by Analog Devices.

Shopping List

This section of the book uses the following components (Please refer to the hardware list file on the RSGB Book File Repository for this book to obtain up-to-date URLs for where to buy these items):

- SAMD21
- ADF5355 Development Board

The ADF5355 is a DDS capable of generating an RF signal from 54MHz all the way up to 13.6GHz. There are now a number of development boards available for this device from the usual outlets, my example is show in '**Figure 14.1**.

A look at the datasheet for this device will soon make you realise that we need some pretty nifty double-precision maths to generate the required register values to program this device.

Figure 14.1 - ADF5355 Development Board

101

Target Freq	2112	MHz	Ref Doubler	No	
Ref	122.88	MHz	Ref Div2	yes	Calc
Chan Space	0.2	MHz	Feedback	Fundamental	
OutDiv	2		Reg0	200220	
PFD	122.88		Reg1	6000001	Enter the Target Frequency, Reference clock multiplier/Dividor selections and click the calc button.
N	34.375		Reg2	C002	
N_Int	34		Reg3	3	
Frac1	6291456		Reg4	3000A784	
Frac1 Int	6291456		Reg5	800025	N will turn red if outside recommended range; change ref doubler or clock divisor to correct.
Frac2	0		Reg6	15220476	
Frac2 Int	0		Reg7	120000E7	
Mod2	3072		Reg8	102D0428	Mod2 will turn red if over 16383
GCD	40		Reg9	302FCC9	
ADCClkDiv	255		Reg10	C03FFA	
			Reg11	61300B	
			Reg12	1041C	

Figure 14.2 - ADF5355 Register Calculator in Excel

I have interpreted the datasheet mathematics into a macro-based Excel spreadsheet which is shown in **Figure 14.2**.

You enter the target frequency and reference clock values into the yellow boxes and click the "Calc" button; the 13 resister values needed to configure the ADF5355 to those requirements are then calculated and displayed in the green boxes.

You will find this Excel spreadsheet in the RSGB book file repository in the directory Arduino Zero. You will also find a simple Arduino sketch for sending the 13 register values to the device over SPI from the SAMD21 MINI development board.

You should have enough knowledge now to take this simple sketch and generate a full-blown signal generator if you wish to do so.

In my case I have a shack based 10MHz frequency reference which is used as the reference clock for an AD4351 which in turn generates a 100MHz reference clock for my AD5355 which has a touch screen display and keypad to generate RF over the full device range.

Things to try

- Have a good look at the spreadsheet macro written in MS Visual Basic – this will help you understand the code needed to generate the register values yourself from a target frequency.
- Check out the work by Andy G4JNT on this topic
- Check out the work by Brian GM8BJF on this topic
- Build yourself a shack signal generator!

The PIC Microcontroller

15

A Programming Introduction

PIC Microcontroller Overview

The PIC Microcontroller family is a collection of MCUs manufactured by Microchip. The first Microchip PICs were available to the market in 1976 when the acronym PIC was defined as Peripheral Interface Controller. By 2013 Microchip had sold more than 12 billion devices and PIC was now defined as Programmable Intelligent Computer. Most programmers incorrectly call PICs Programmable Interface Controller (I don't know why either).

The Microchip PIC family contains many devices including 8-bit, 16-bit and 32-bit MCUs ranging from simplistic 8 pin DIL devices to 144 pin SMD chips. On-board peripherals vary by device but may include digital I/O, ADC and DAC modules, communication interfaces such as UART, I2C, SPI, CAN and USB. Low power and very high-speed devices also exist within the range.

PIC devices are popular with both industrial developers and hobbyists due to their low cost, wide availability, large user base, extensive collection of application notes, availability of free development tools, serial programming and re-programmable flash-memory capabilities.

Microchip now also own the AVR range of MCUs, having acquired the technology in 2016 from Amtel.

PIC Selection

For our experiments we will use one of the 8-bit MCUs from the PIC range; do not be confused by the naming, PIC18, PIC16, PIC12 and PIC10 series MCUs are all 8-bit devices.

We will use the Microchip development board DM164142 which is low-cost and contains a PIC16F18877 MCU requiring no external hardware programmer.

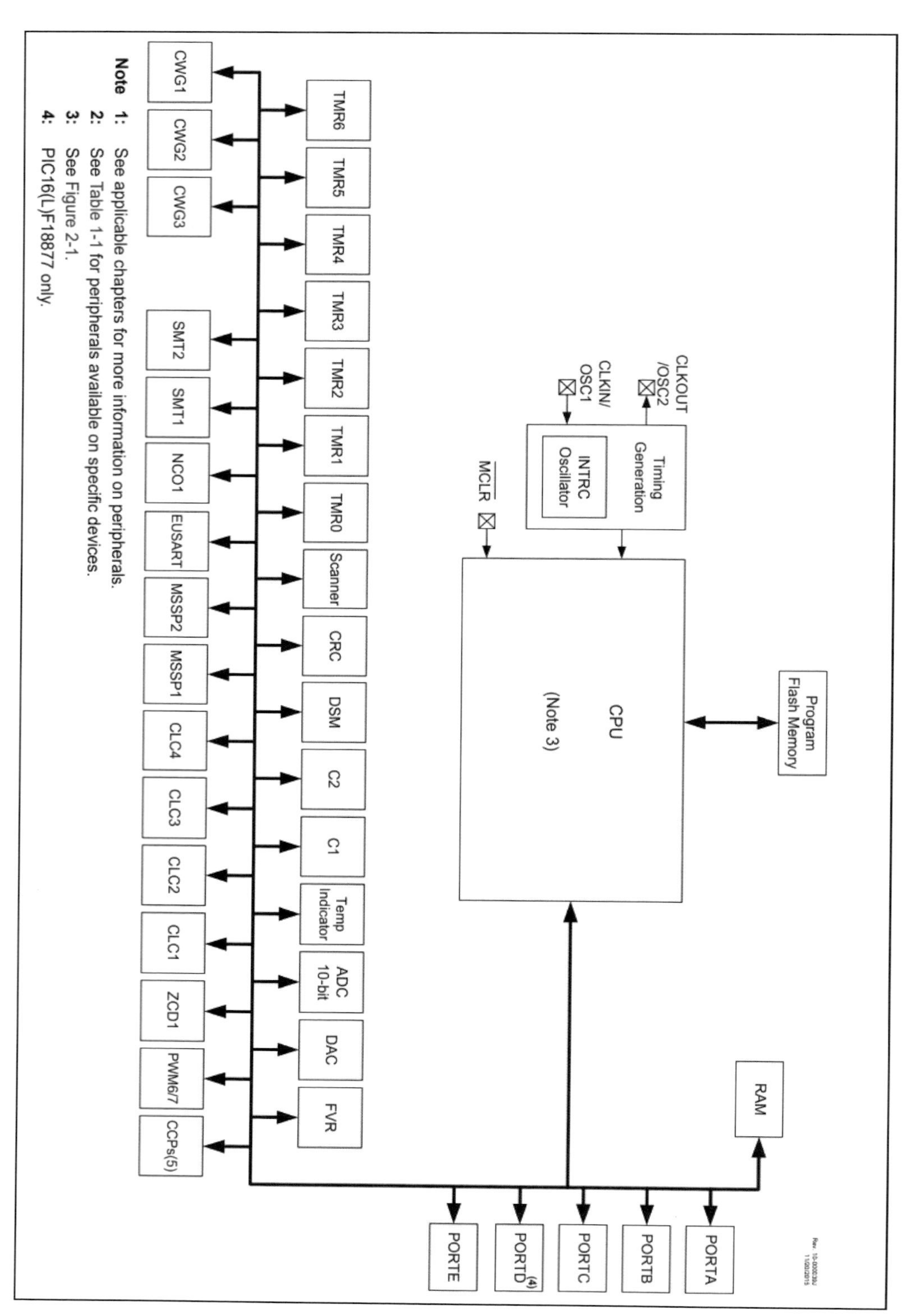

Figure 15.1 - PIC16F18877 Block Diagram

The datasheet for the PIC16F18877 MCU is available from the Microchip website and contained in the RSGB book file repository in the directory PIC.

Some of the main features of the PIC16F18877:

- C-Compiler optimised
- Up to 32MHz clock input
- 3 8-bit timers
- 4 16-bit timers
- Up to 56kB Flash program memory
- 4k Data SRAM
- 256B EEPROM
- Power saving features
- Comprehensive set of both analogue and Digital Peripherals
- 10-bit ADC
- Flexible oscillator structure

A block diagram of the PIC16F18877 is shown in **Figure 15.1** and is reproduced from the device datasheet.

You can see the vast array of peripherals available on the PIC16F18877 device; do not be fooled by the low cost, this is an extremely powerful MCU.

Shopping List

This section of the book uses the following components (Please refer to the hardware list file on the RSGB Book File Repository for this book to obtain up-to-date URLs for where to buy these items):

- DM164142 development board
- 7-segment LED (common anode or cathode)
- 470-ohm resistors (up to 14)
- Tactile switch
- Header pins
- MCP3021
- LCD display
- 10k pre-set potentiometer

Development Environment

The DM164142 development board is designed to be used with the Microchip MPLAB Xpress® which is a cloud-based IDE. This means that you do not install any software on your own PC but access the IDE and storage is via a web browser.

The development board also supports the more traditional PC hosted IDE from Microchip called MPLAB®. It is this route that we will go with our PIC experiments. The MPLAB® development environment is combined with the XC8© compiler when developing C code for the 8-bit MCU family.

These software tools combined with our board create a complete development environment.

Note the XC8® compiler operates in one of three modes: Free, Standard and Pro. Only the Free option is available without purchasing a licence, but this mode of operation is normally more than adequate for hobby usage. The modes primarily differ by the level of optimisation performed by the compiler, higher levels of optimisation are needed where execution speed or code size are critical.

Software Installation

MPLAB X IDE ®

Download the latest release of the software from the Microchip website and follow the instructions for installation. On Windows this consists of downloading the executable installer and following the on-screen instructions. At the time of writing the MPLAB X IDE ® was available from:

https://www.microchip.com/en-us/development-tools-tools-and-software/mplab-x-ide

be warned, this is a large download!

XC8 ® C-Compiler

Download the latest version of the XC8 ® C-Compiler installation package from the Microchip website. Make sure to select the Free installation option so that no additional licence is required.

At the time of writing the MPLAB XC8 ® C-Compiler for Windows was available from:
https://www.microchip.com/mplabxc8windows

MPLAB ® Code Configurator

Start the MPLAB X IDE ® and use the Tools > Plugins menu option. From the "Available Plugins" tab you should be able to find 'MPLAB ® Code Configurator'. Select that option and then click install and follow the on-screen instructions.

Preparing the Hardware

The DM164142 development board needs soldered headers to enable the board to be inserted into a breadboard. We will then connect a LED and 1k series resistor between pin RB0 and ground. The anode of the LED connects directly to RB0 with the 1k resistor connected to ground and the cathode of the LED.

Once you have completed your hardware preparation, connect the DM164142 development board to your PC using a USB cable. Once connected to your PC you will find an additional "flash drive" available on the PC. My board creates a drive on the PC called 'XPRESS'.

Copy the file LEDblink.hex from the RSGB book file repository in the Directory: PIC into this flash drive and it will be automatically programmed into the PIC on the development board. You have now executed your first PIC embedded program – congratulations.

A word about device configuration

When we looked at STM32 programming using the STM Cube IDE, we touched upon the concept of device configuration. We also understood that a large amount of low-level configuration is being done for us in the background when we use an IDE like Arduino.

Device configuration in the world of PIC programming is nothing other than critical. If you don't get your device configuration right, your software isn't going to work, or is going to produce very unexpected results.

There are primarily three parts to device configuration in a PIC MCU:

- Configuration words
- Code protection
- Device ID

The configuration words define things like the oscillator selection, memory usage et cetera. In the PIC16F18877 there are five configuration registers which need to be set according to the desired oscillator and other parameters required.

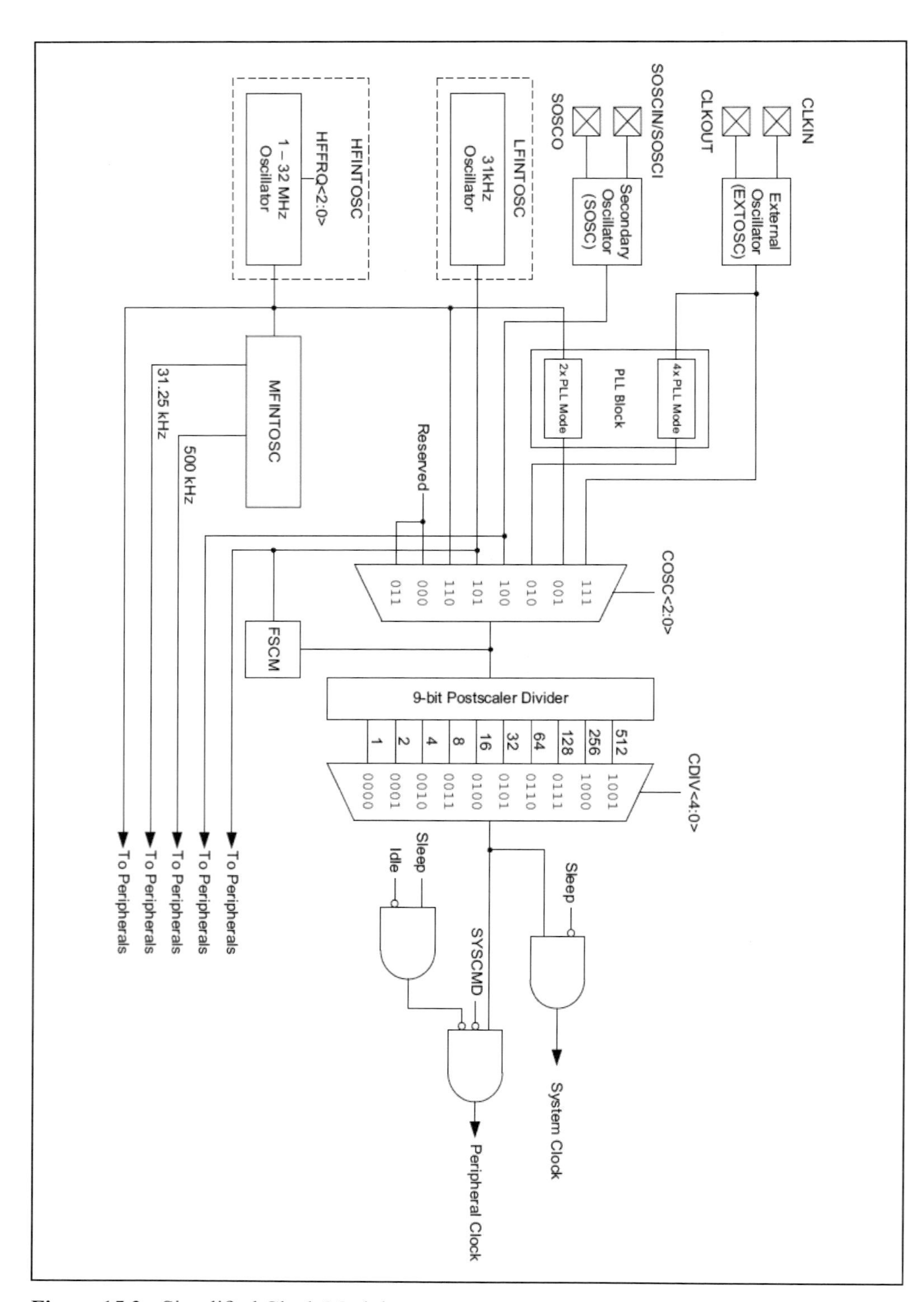

Figure 15.2 - Simplified Clock Module

Configuration register 1

Configuration register 1 deals with the oscillator configuration.

The oscillator on the PIC16F18877 can be configured as internal or external and when external one of:

- External Clock Low Power (below 500kHz) – ECL
- External Clock Medium Power (500kHz to 8MHz) – ECM
- External Clock High Power (above 8MHz) – ECH
- Low Power Crystal Mode – LP
- Medium Gain Crystal or Ceramic Resonator (between 100kHz and 4 MHz) – XT
- High Gain Crystal or Ceramic Resonator (above 4MHz) - HT

There are two internal oscillators available in the PIC16F18877 MCU, these are:

- High Frequency internal oscillator – software configurable up to 32MHz.
- Low Frequency internal oscillator – factory calibrated at 31kHz.

A simplified block diagram of the clock module is shown in **Figure 15.2**.

If you followed the clock configuration of the STM32 MCU, this concept should be familiar to you; clock selection from one of internal or external, PLL multipliers and other configuration set the system and peripheral clocks.

Note the concept of a secondary oscillator is present on the PIC16F18877 and many other PICs. The secondary oscillator would generally be a low power quartz crystal oscillator used when the software is in a hibernated or low power operating state; this secondary oscillator has to be selected by software.

Configuration Register 2

Configuration register 2 deals with supervisory options for the device. These configuration items include enabling of background debugging, reset on stack over/underflow, brown-out reset levels, power up timers et cetera.

Configuration Register 3

Configuration register 3 is the setup of the PIC watchdog timer.

Configuration Register 4

Configuration register 4 enables the selection of low voltage programming, scanner module and the write protection of the FLASH memory.

Configuration Register 5

Configuration register 5 selects the protection or otherwise of the code within the PIC. If you prize your embedded programming software be sure to protect it in the device so that it cannot be read and disassembled back to machine code.

Back to Blinky Again

Whilst the details of the configuration registers may sound a little complex, the setting up of this device configuration can be made quite easy for us. Let's put all the theory of the PIC we have so far into practice and develop our first embedded PIC program from scratch – let's do Blinky on our PIC.

We should now have our PIC hardware setup on a breadboard with an LED connected to pin RB0. We also have our PC with MPLAB X IDE ®, XC8 ® Compiler and the MPLAB ® Code Configurator.

Launch the MPLAB X IDE on your PC and select file – new project, use the defaults of "Microchip Embedded" for the Category and "Standalone Project" for the Project Type as shown in **Figure 15.3**.

Click next then you will be presented with device selection.

You can type into the device name box or produce a sub-set of devices to choose from by selecting a family. Our device is the PIC16F18877 and is in the family "baseline 8-bit MCU". It can be selected by simply typing the device name into the appropriate box as shown in **Figure 15.4**.

As our development board is programmed by copying the IDE generated hex file to a PC flash drive, we do not make any tool selection.

Click Next and you will be presented with a selection of compilers to choose from, in our case we need to choose the XC8 compiler we installed previously, so select that and click next once more.

Now give the project an appropriate name, I like to put the device first, so I would use "PIC16F18877 Blinky" as my project name. Select or create an appropriate directory (I used PIC here) and then click Finish.

MPLAB X will now create a set of folders and files accessible in the project structure within the tool.

Now use the Tools -> Embedded -> MPLAB Code Configurator menu option to start the code configuration tool. Alternatively, you can click on the MCC button on the toolbar:

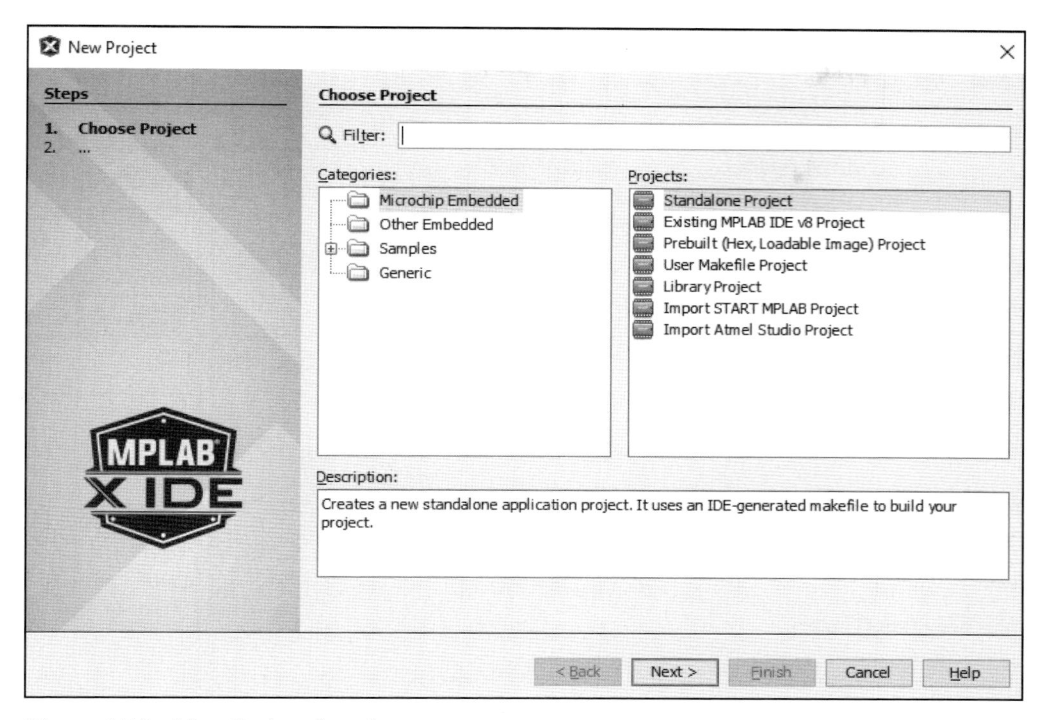

Figure 15.3 - New Project Step 1

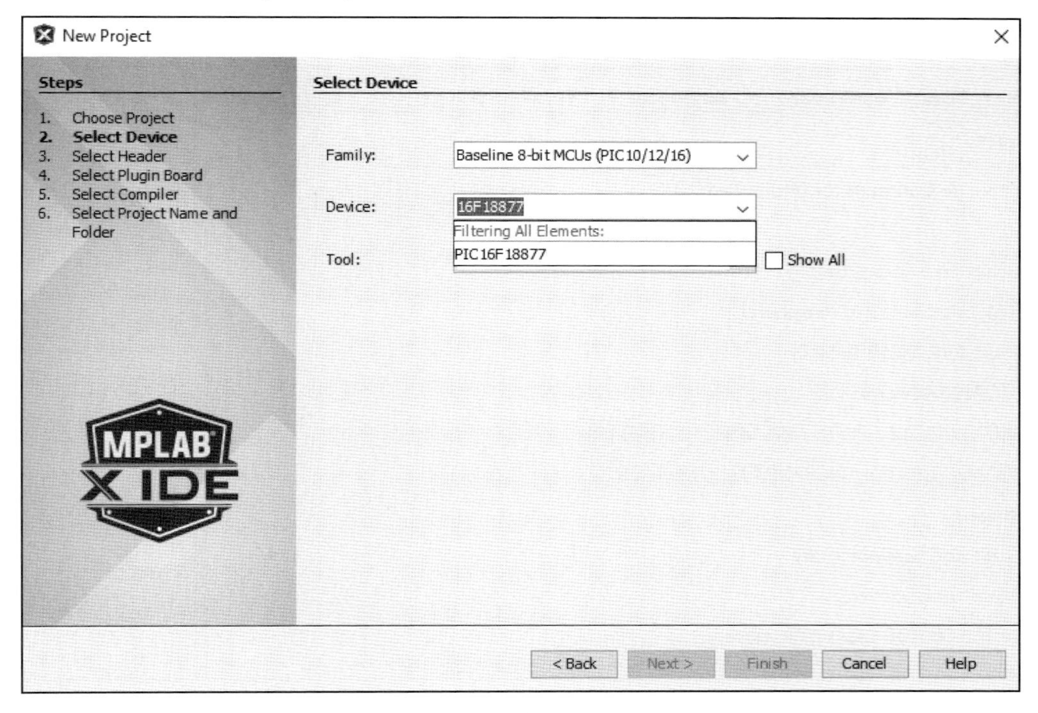

Figure 15.4 – New Project - PIC Device Selection

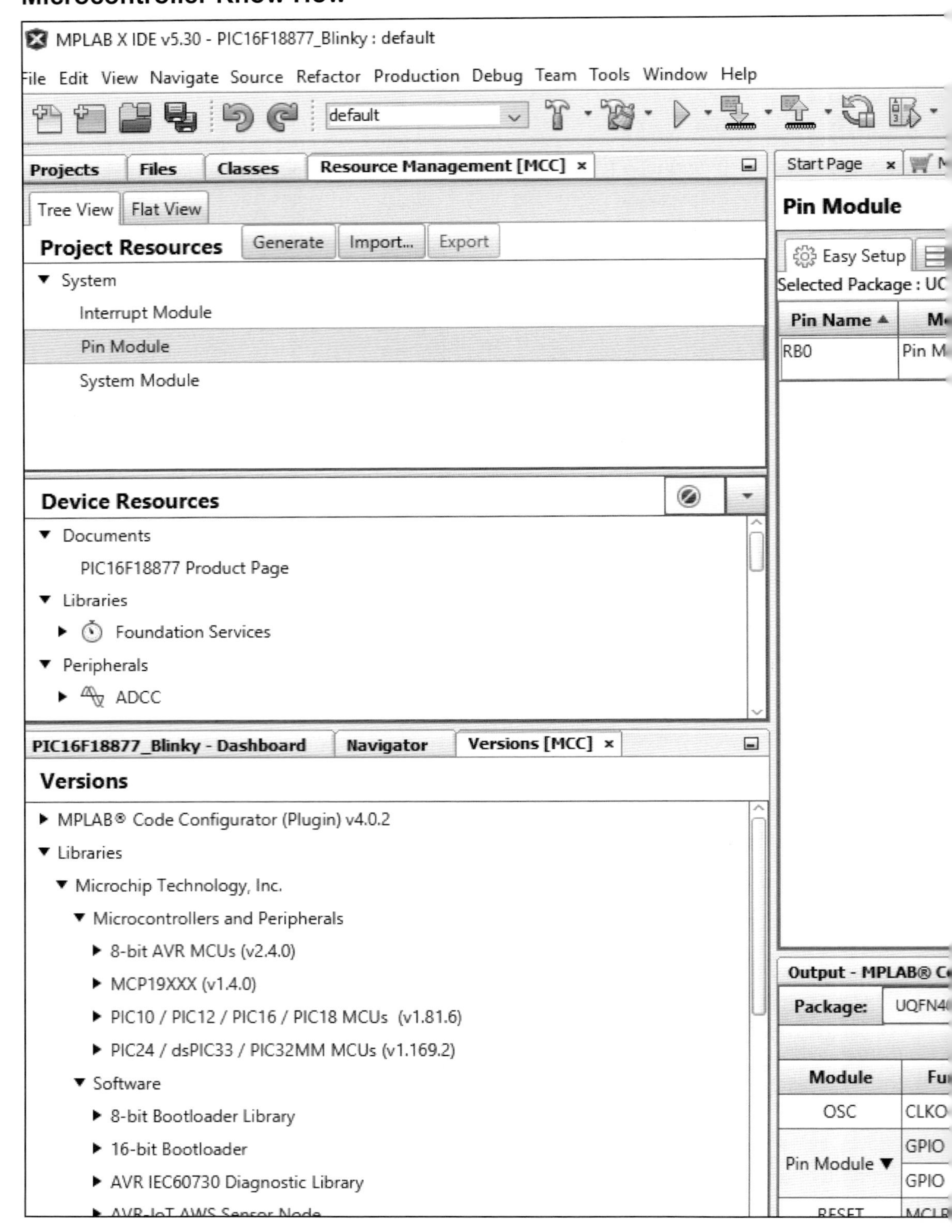

Figure 15. 5 - RB0 pin Custom Name assignment

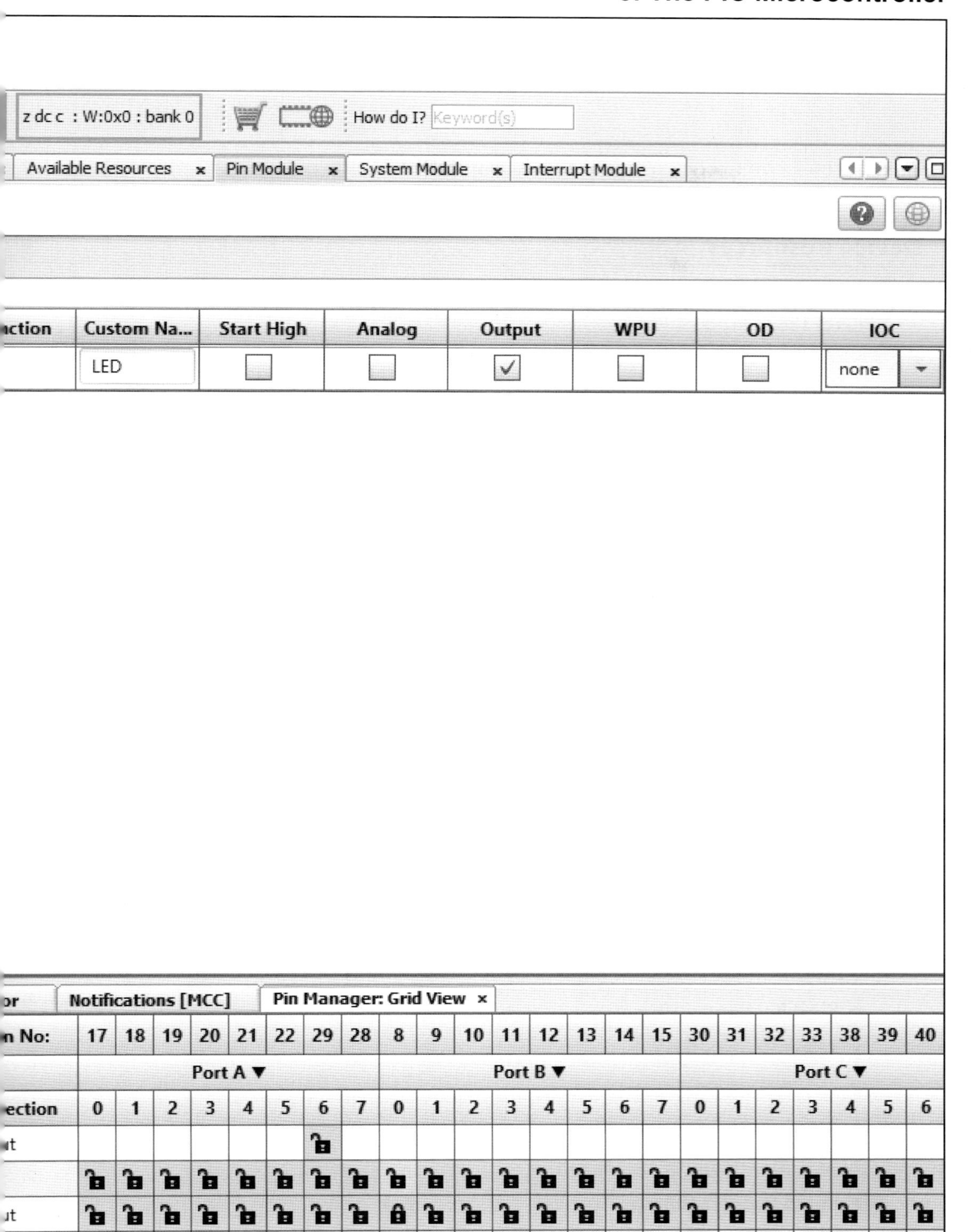
z dc c : W:0x0 : bank 0
How do I?
Keyword(s)
Available Resources
Pin Module
System Module
Interrupt Module
Custom Na...
Start High
Analog
Output
WPU
OD
IOC
LED
none
Notifications [MCC]
Pin Manager: Grid View
Port A
Port B
Port C

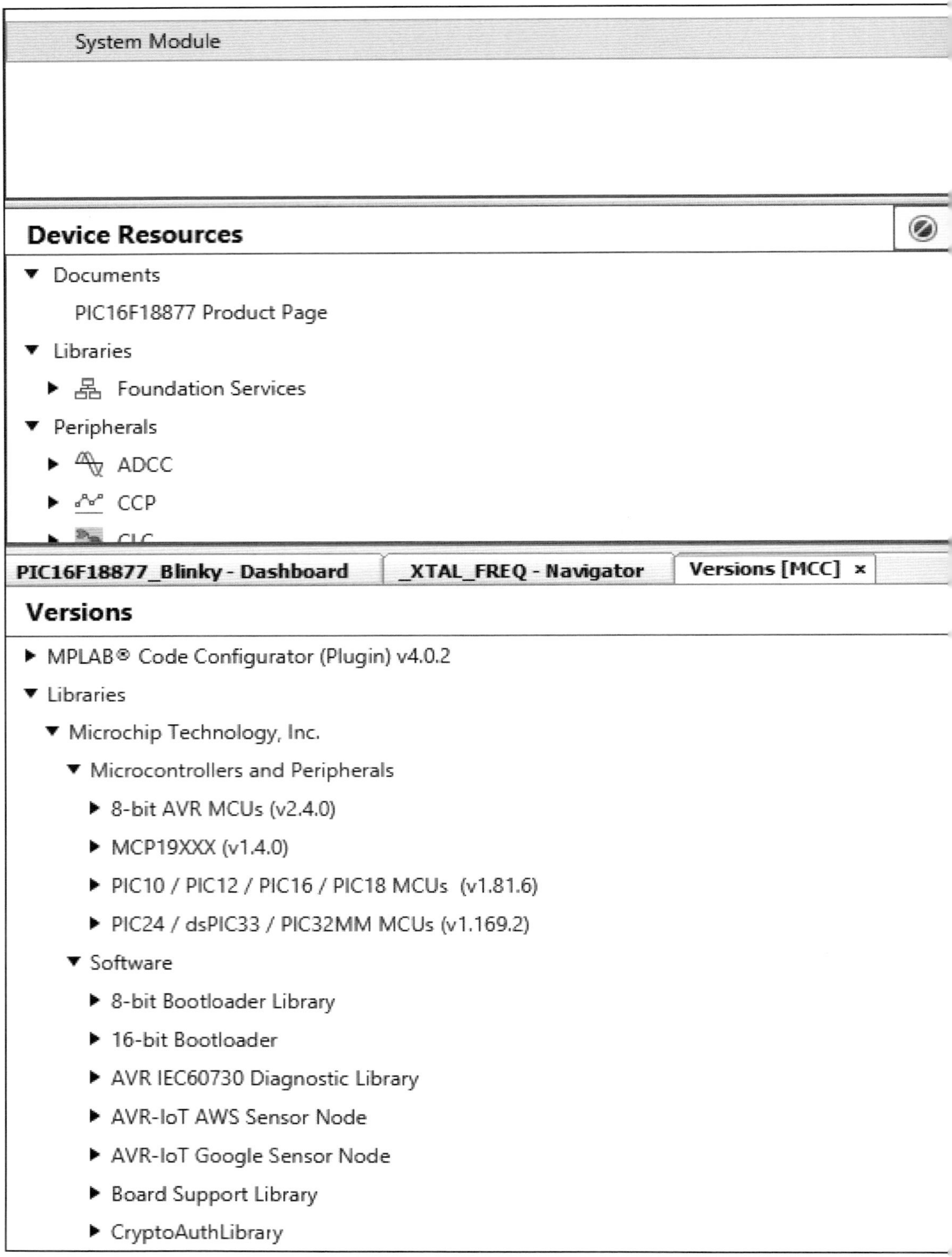

System Module

Device Resources

- Documents
  - PIC16F18877 Product Page
- Libraries
  - Foundation Services
- Peripherals
  - ADCC
  - CCP
  - CLC

PIC16F18877_Blinky - Dashboard | _XTAL_FREQ - Navigator | Versions [MCC] ×

Versions

- MPLAB® Code Configurator (Plugin) v4.0.2
- Libraries
  - Microchip Technology, Inc.
    - Microcontrollers and Peripherals
      - 8-bit AVR MCUs (v2.4.0)
      - MCP19XXX (v1.4.0)
      - PIC10 / PIC12 / PIC16 / PIC18 MCUs (v1.81.6)
      - PIC24 / dsPIC33 / PIC32MM MCUs (v1.169.2)
    - Software
      - 8-bit Bootloader Library
      - 16-bit Bootloader
      - AVR IEC60730 Diagnostic Library
      - AVR-IoT AWS Sensor Node
      - AVR-IoT Google Sensor Node
      - Board Support Library
      - CryptoAuthLibrary

Figure 15.6 - Code Generator Example (left)

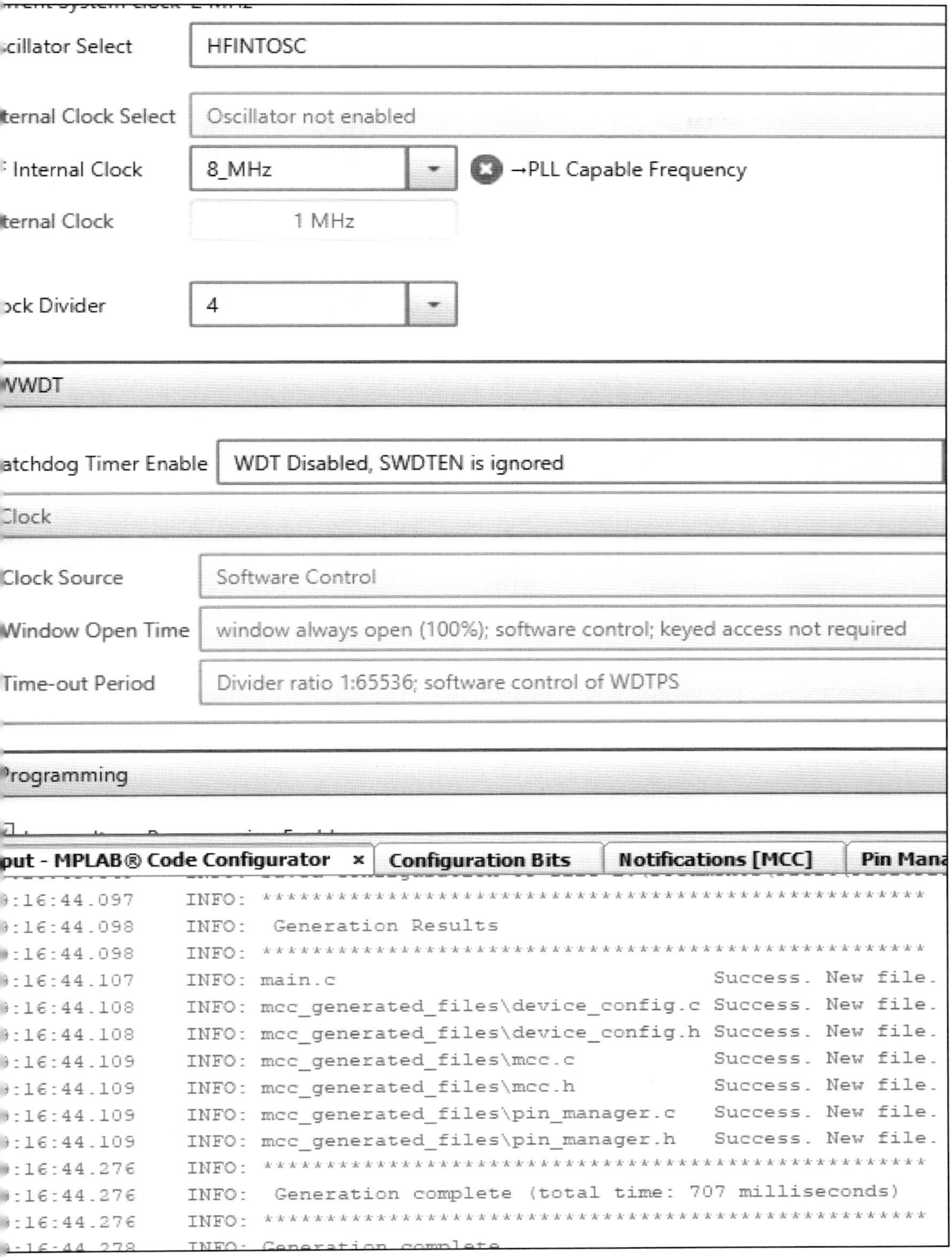

:cillator Select	HFINTOSC
:ternal Clock Select	Oscillator not enabled
: Internal Clock	8_MHz ▾ ⊗ →PLL Capable Frequency
:ternal Clock	1 MHz
:ock Divider	4 ▾

WWDT

atchdog Timer Enable	WDT Disabled, SWDTEN is ignored

Clock

Clock Source	Software Control
Window Open Time	window always open (100%); software control; keyed access not required
Time-out Period	Divider ratio 1:65536; software control of WDTPS

Programming

put - MPLAB® Code Configurator × | Configuration Bits | Notifications [MCC] | Pin Mana

```
:16:44.097   INFO: ***************************************************
:16:44.098   INFO:  Generation Results
:16:44.098   INFO: ***************************************************
:16:44.107   INFO: main.c                                   Success. New file.
:16:44.108   INFO: mcc_generated_files\device_config.c Success. New file.
:16:44.108   INFO: mcc_generated_files\device_config.h Success. New file.
:16:44.109   INFO: mcc_generated_files\mcc.c            Success. New file.
:16:44.109   INFO: mcc_generated_files\mcc.h            Success. New file.
:16:44.109   INFO: mcc_generated_files\pin_manager.c    Success. New file.
:16:44.109   INFO: mcc_generated_files\pin_manager.h    Success. New file.
:16:44.276   INFO: ***************************************************
:16:44.276   INFO:  Generation complete (total time: 707 milliseconds)
:16:44.276   INFO: ***************************************************
```

ure 15.6 - Code Generator Example (right)

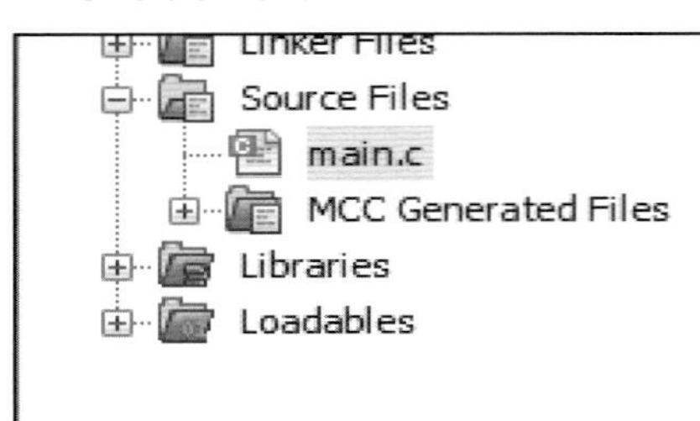
Linker Files
Source Files
main.c
MCC Generated Files
Libraries
Loadables

PIC16F18877_Blinky - Dashboard
main() - Navigator
Versions [MCC] ×
Versions
MPLAB® Code Configurator (Plugin) v4.0.2
Libraries
Microchip Technology, Inc.
Microcontrollers and Peripherals
8-bit AVR MCUs (v2.4.0)
MCP19XXX (v1.4.0)
PIC10 / PIC12 / PIC16 / PIC18 MCUs (v1.81.6)
PIC24 / dsPIC33 / PIC32MM MCUs (v1.169.2)
Software
8-bit Bootloader Library
16-bit Bootloader
AVR IEC60730 Diagnostic Library
AVR-IoT AWS Sensor Node
AVR-IoT Google Sensor Node
Board Support Library
CryptoAuthLibrary

Figure 15.7 – Editing main.c (left)

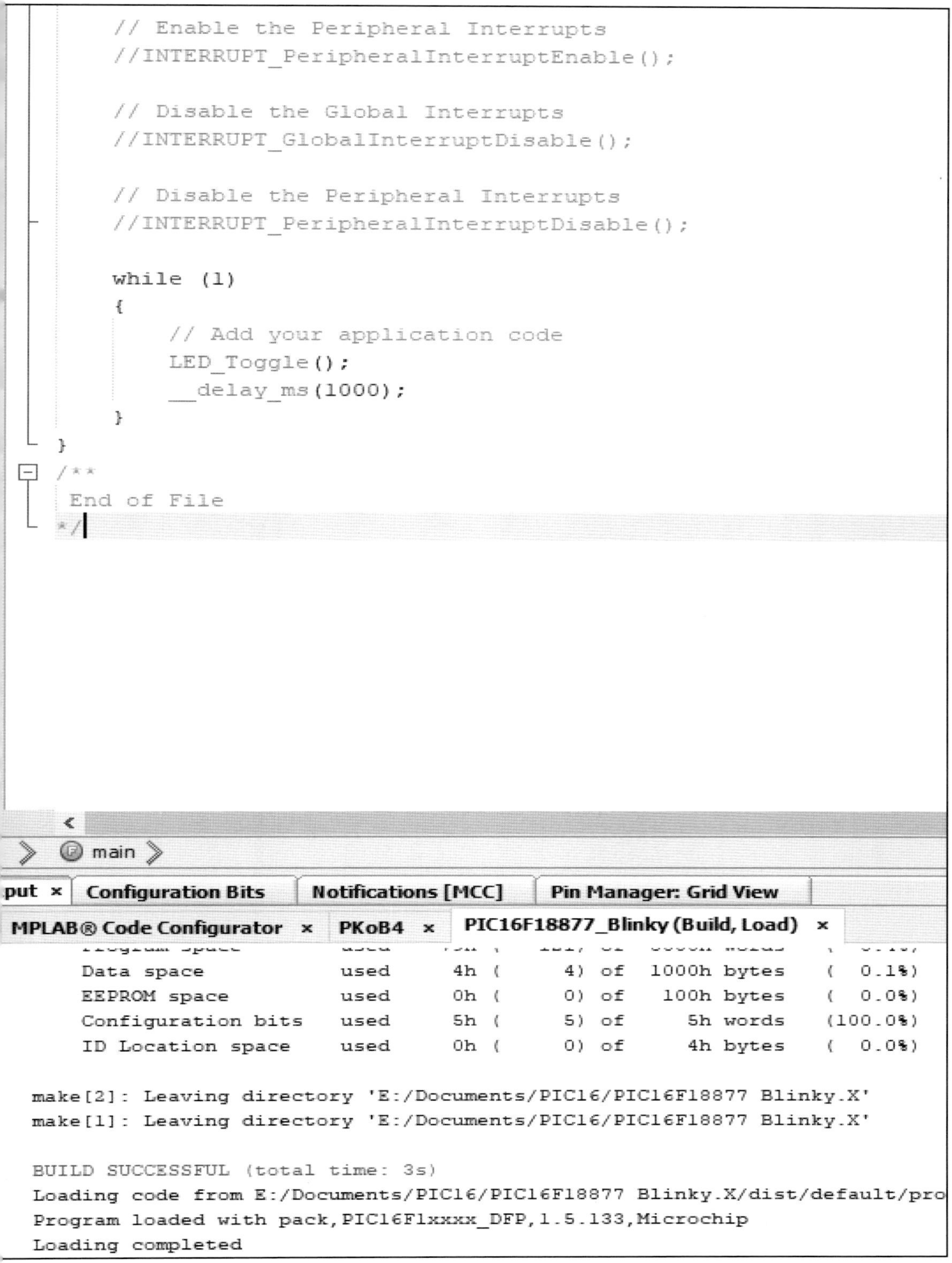

```
    // Enable the Peripheral Interrupts
    //INTERRUPT_PeripheralInterruptEnable();

    // Disable the Global Interrupts
    //INTERRUPT_GlobalInterruptDisable();

    // Disable the Peripheral Interrupts
    //INTERRUPT_PeripheralInterruptDisable();

    while (1)
    {
        // Add your application code
        LED_Toggle();
        __delay_ms(1000);
    }
}
/**
 End of File
*/
```

```
main
put × | Configuration Bits | Notifications [MCC] | Pin Manager: Grid View
MPLAB® Code Configurator × | PKoB4 × | PIC16F18877_Blinky (Build, Load) ×

    Data space               used    4h (     4) of  1000h bytes   (  0.1%)
    EEPROM space             used    0h (     0) of   100h bytes   (  0.0%)
    Configuration bits       used    5h (     5) of     5h words   (100.0%)
    ID Location space        used    0h (     0) of     4h bytes   (  0.0%)

make[2]: Leaving directory 'E:/Documents/PIC16/PIC16F18877 Blinky.X'
make[1]: Leaving directory 'E:/Documents/PIC16/PIC16F18877 Blinky.X'

BUILD SUCCESSFUL (total time: 3s)
Loading code from E:/Documents/PIC16/PIC16F18877 Blinky.X/dist/default/pro
Program loaded with pack,PIC16F1xxxx_DFP,1.5.133,Microchip
Loading completed
```

ıre 15.7 – Editing main.c (right)

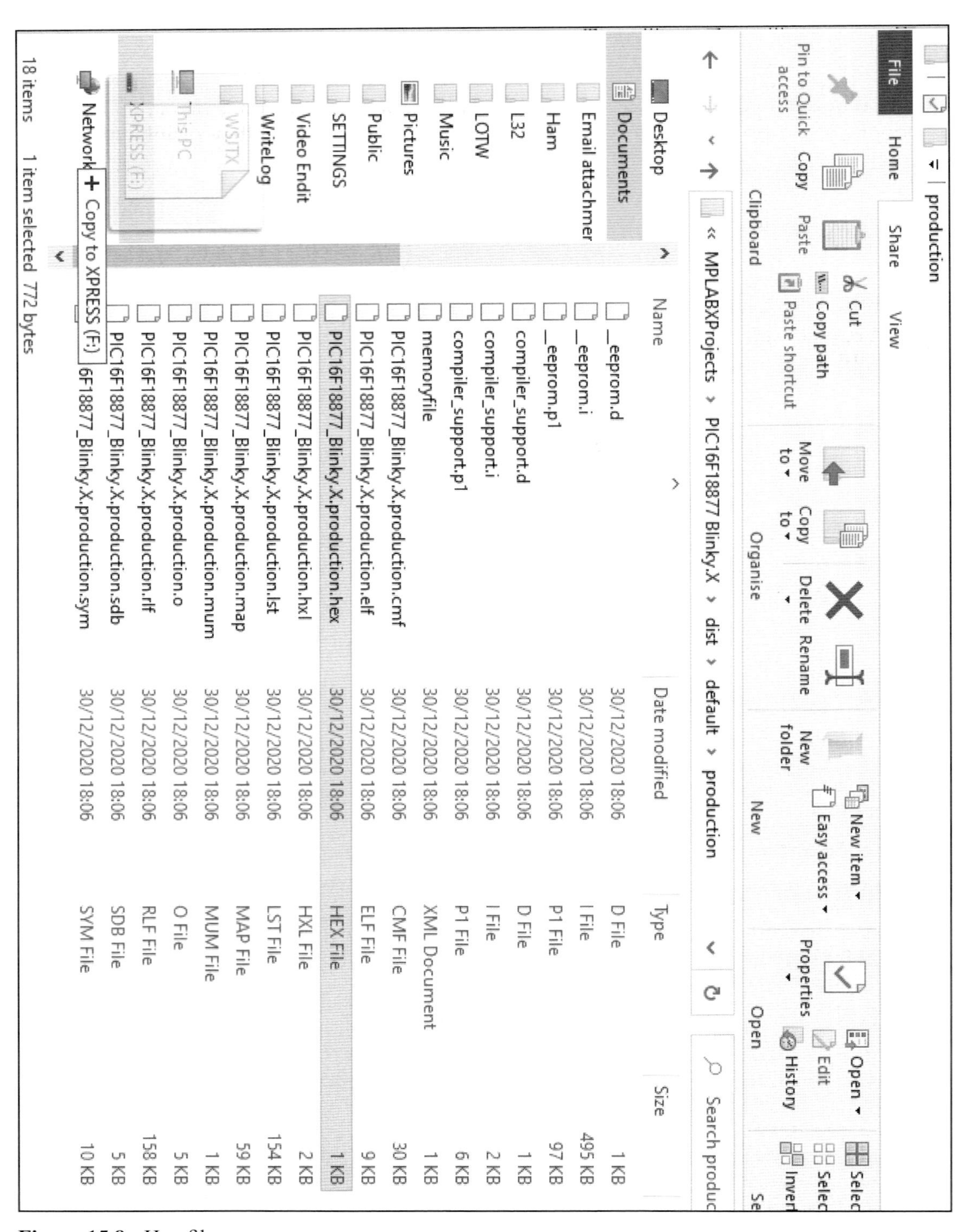

Figure 15.8 - Hex file copy process

Once again if you followed the clock configuration of the STM32 MCU, this concept should be familiar to you as we are now presented with a graphical representation of the MCU device and can select the configuration options we wish to use.

We can also right click on the individual pins of the MCU shown in the graphical representation and select from the options available.

In our case, we need to set pin RB0 as a GPIO (General Purpose Input Output) Output. Once you have done that click on the Pin Module option under Project Resources and give the RB0 pin a Custom Name of "LED" as shown in **Figure 15.5**. We also need to deselect the Analog option for the pin.

Now click on System Module under Project resources and under the Easy Setup tab select an appropriate oscillator configuration. I have used the High Frequency Internal Oscillator (HFINTOSC) with a clock frequency of 8MHz and a Clock divider of 4 giving an actual oscillator frequency of 2MHz.

The defaults for other options should work for us so now click the generate button next to Project Resources and the code will auto generate as shown in **Figure 15.6**.

Now go to the Projects tab and find the source files section for our project and open main.c as shown in **Figure 15.7**.

We are going to add the following lines below the "//Add your application code" comment:

```
LED_Toggle();
__delay_ms(500);
```

To complete the project, we will now click on Build icon:

Find the hex file that has been generated by the build process, located in:

<<source directory>>\<<project name>>\dist\default\production

By default, this would be:

<<User Area>>\<<user>>\MPLABXProjects\PIC16F18877 Blinky.X\dist\default\production

To programme our development board, we need to copy the hex file into the associated flash drive.

This action is shown in **Figure 15.8**.

Once the file is copied, the code should now automatically program into our DM164142 development board.

Congratulations, you have just successfully completed your first embedded PIC project.

RSGB BOOKSHOP

Always the best Amateur Radio books

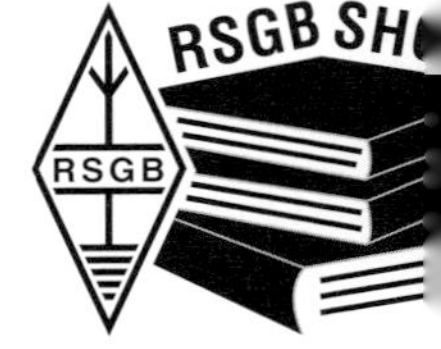

Software Defined Radio

By Andrew Barron, ZL3DW

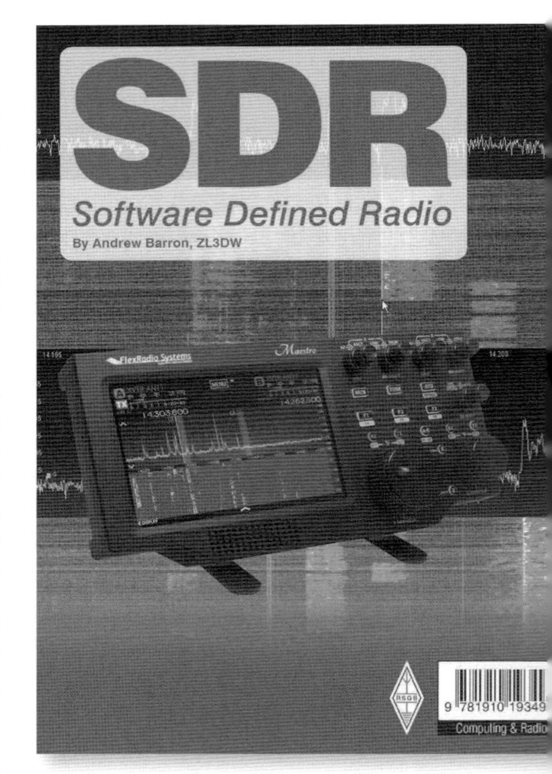

Everyone is talking about software defined radio (SDR) but is SDR right for you? Software Defined Radio sets out to explain the basics without getting to technical and is written to help you too get the most out of your SDR. It will even help you decide what to buy.

Written by New Zealand based and acknowledged SDR expert Andrew Barron, ZL3DW, Software Defined Radio covers a huge range of material. The use of SDR by radio amateurs is growing rapidly in popularity as they become aware of the great features and performance on offer. Not only does this book cover how SDR works there are details the different types of software that are available, what is different about them and even what is better. There is a wealth of useful information included and even guides to what to look for when you are buying equipment. There are guides to using SDR with CW, Digital Modes, Contesting, EME, Microwaves, Satellites and much more. You will find information on over 60 SDR radios that you can buy today featuring leading brands such as FlexRadio, Elecraft, Anan, Expert, Elad, Icom, WiNRADiO, SDRplay, FUNcube and many more.

Software Defined Radio is intended for radio amateurs, short wave listeners or anyone interested in radio technology. If you are interested in the technology of what was once, the domain of a few dedicatec hackers and experimenters, the future of this exciting and fast developing area of radio or simply wan to buy a SDR radio, this book is thoroughly recommended reading.

Size 174x240mm, 304 pages
ISBN: 9781 9101 9349 5
Price £14.99

Don't forget RSGB Members always get a discount

Radio Society of Great Britain www.rsgbshop.org

3 Abbey Court, Priory Business Park, Bedford, MK44 3WH. Tel: 01234 832 700 Fax: 01234 831 496

FROM **FREE P&P**
on orders over £30. See T&

16

Notes on PIC GPIO

In our selected PIC we have a total of 5 ports containing externally connectable pins. The ports are rather originally called PORTA, B, C, D and E.

Ports A to D on our device are 8-bit ports, Port E only has 4-bits. We therefore have a total of 36 possible GPIO pins on the device. However, the pins will often be shared with other peripherals and can only be used as GPIO pins when the shared peripheral is not in use.

Section 12 of the device datasheet contains all the detail on the GPIO ports.

Each pin of each port can be represented by the generic block diagram shown in **Figure 16.1**. Each port register will be 8 bits (except PORTE) and each bit within the register will represent an individual GPIO pin. PORTA bit 0 registers are associated with pin RA0, PORTB bit 1 with RB1, et cetera.

Each GPIO port has a set of associated registers, for PORTx these are:

- PORTx – reads the voltage levels on the pins of the device
- LATx – reads the status of the output latch
- TRISx – configures the data direction
- ANSELx – sets the pin as analogue
- WPUx – selects weak pull-up
- INLVLx – Input Level Control
- SLRCONx – Slew Rate Control
- ODCON – Open Drain Selection

When data is written by the MCU to either the PORTx or LATx register, the data value gets stored in the data register. If the TRISx register is enabled

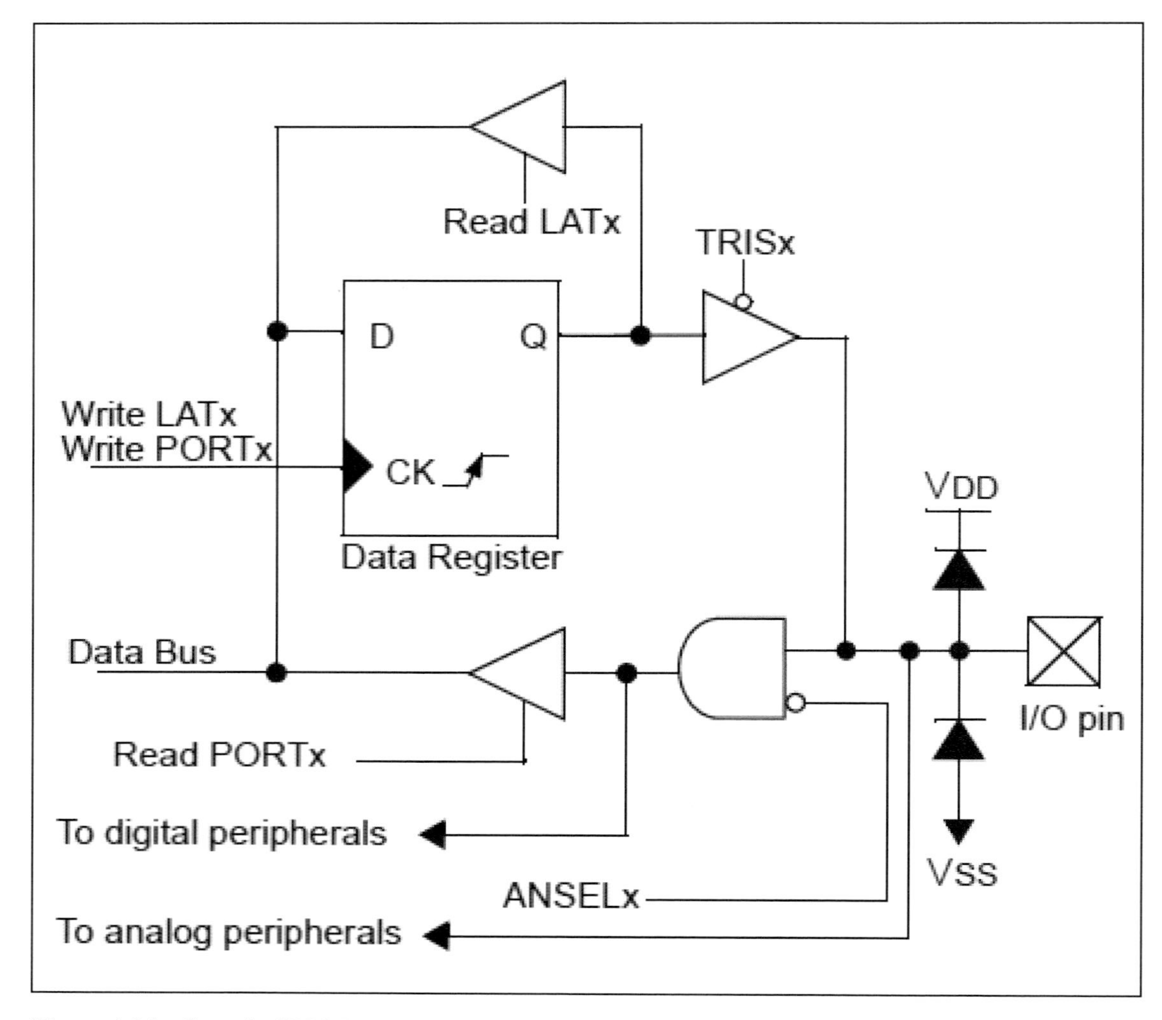

Figure 16.1 - Generic GPIO Port

(selecting output by writing 0 (zero)), the data value will be assigned to the physical pin of the port.

When data is read from the PORTx register, the actual value on the pin is read by the MCU as long as the TRISx register is disabled (selecting input by writing 1).

Therefore, to configure a pin as input we write 0 (zero) to the corresponding TRIS register and to configure as output, we write a 1.

Good practice suggests that writing to a port should be done using the LATx register and reading should be directly from the PORTx register. Writing to the LATx register avoids a problem in MCU operation known as read-modify-write.

The ANSELx register enables the pin routing to the ADC peripheral and there-

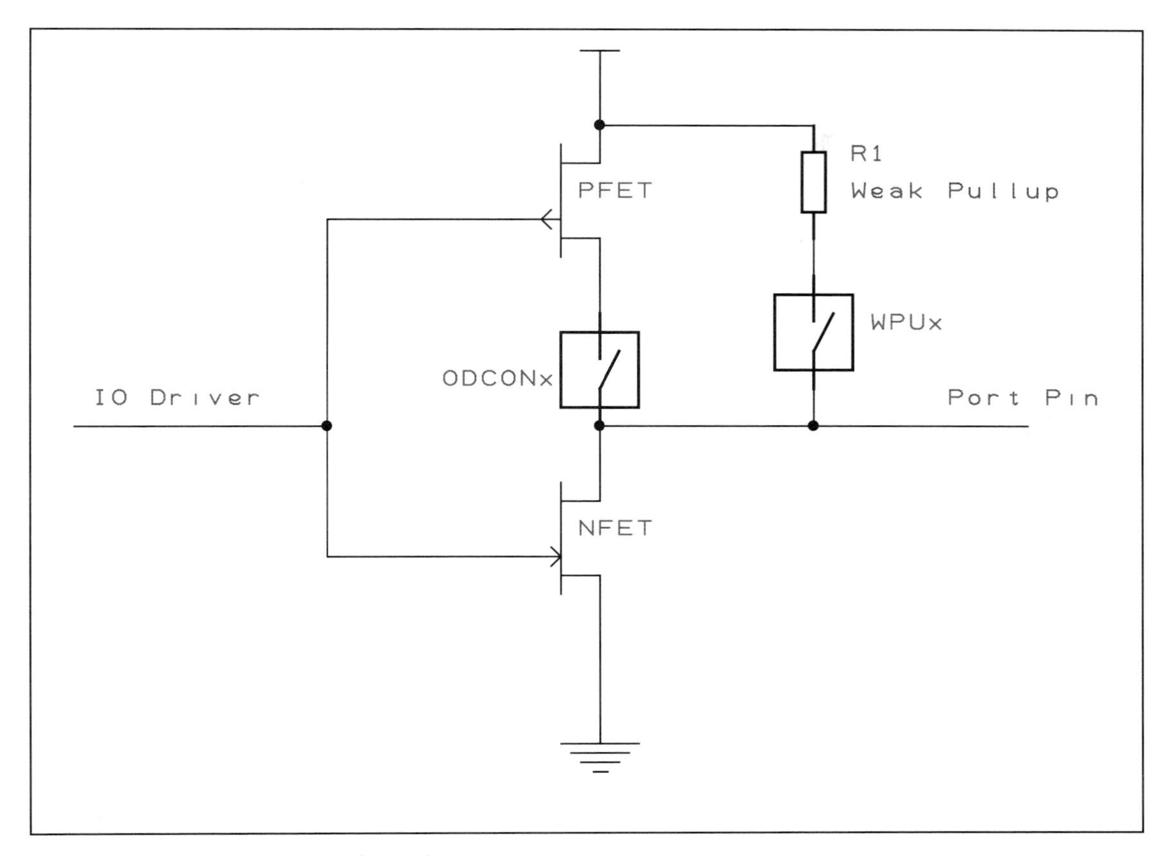

Figure 16.2 - Port Output Configuration

fore it is important to disable the ANSELx register when using pins for digital IO.

In the same way as we found on the STM32, there are also two possible configurations of digital outputs on the PIC, either open-drain or a push pull. Open drain is useful when multiple gates or pins are connected together with a pull-up resistor. If all the pins are high, they are all open circuit and the pull-up drives all the pins high. If any pin is low they all go low as they are tied together. This effectively creates gated output sets. The output pin configuration is illustrated in **Figure 16.2**. Q1 is a p-MOSFET and Q2 is an n-MOSFET.

If the SLRCONx register is set, the corresponding drive is slew rate (the rate at which the signal changes) limited, when cleared the pin is driven at the maximum slew rate possible.

The WPUx registers enable internal pull-ups on the corresponding pins.

The internal diodes connected to the pin provide over-voltage and polarity protection to the internals of the MCU.

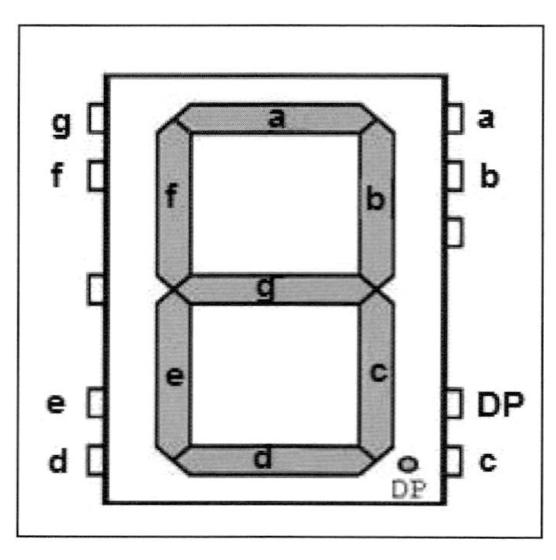

Figure 16.3 - Typical 7-segment LED

Controlling 7-segment LEDs

Introduction

We are going to use a 7-segment display as an exercise in GPIO port programming.

Hardware Setup

The 8-bit output register in the PIC is ideal for controlling a 7-segment LED display. The segments of a 7-segment LED are typically labelled A to E plus a decimal point. A typical 7-segment LED pinout is shown in **Figure 16.3**.

We have learned that the 8 GPIO pins are represented by individual bits within a single register, therefore we can easily construct a table of register values that will represent each digit of the 7-segment display. The logic and associated 8-bit HEX value is shown in **Figure 16.4** for both Common Anode and Common Cathode displays.

	Segment											
Digit	a	b	c	d	e	f	g	DP	Common Cathode		Common Anode	
0	1	1	1	1	1	1	0	0	F	C	0	3
1	0	1	1	0	0	0	0	0	6	0	9	F
2	1	1	0	1	1	0	1	0	D	A	2	5
3	1	1	1	1	0	0	1	0	F	2	0	D
4	0	1	1	0	0	1	1	0	6	6	9	9
5	1	0	1	1	0	1	1	0	B	6	4	9
6	1	0	1	1	1	1	1	0	B	E	4	1
7	1	1	1	0	0	0	0	0	E	0	1	F
8	1	1	1	1	1	1	1	0	F	E	0	1
9	1	1	1	1	0	1	1	0	F	6	0	9
DP	0	0	0	0	0	0	0	1	0	1	F	E
	Rx7	Rx6	Rx5	Rx4	Rx3	Rx2	Rx1	Rx0				
	Pin											

Figure 16.4 - 7-Segment Display Values

PORT Pin	7-Segment
RA7	A
RA6	B
RA5	C
RA4	D
RA3	E
RA2	F
RA1	G
RA0	DP

Table 1

Assuming we are going to use PORTA for our 7-segment display, we would wire the display connections as follows, each one with a current limiting resistor: see **Table 1**

Then, for example, to display digit 0, we need to light segments a, b, c, d, e and f of the 7-segment display, if this is a common cathode device we would write hex FC to the PORT register setting pins RA7, RA6, RA5, RA4, RA3 and RA2 of the port to 1 and pins RA1 and RA0 to 0, if a common anode device, hex 03 which would invert our output.

Software

Now we will create a new project for our 7-segment display control. Use File->New Project and select Microchip Embedded and Standalone as we did last time.

Once again select our device as the PIC16F18877 and leave the tool selection blank or set to no tool.

Select the XC8 compiler and give the project an appropriate name and click Finish.

Instead of using the Code Generator, let's setup our configuration registers in a different way; select Window -> Target Memory Views -> Configuration Bits. We need to make three changes from the defaults presented:

- In the CONFIG1 register, set FEXTOSC to OFF (selects the internal oscillator)
- In the CONFIG1 register, set RSTOSC to HFINT1 (sets a 1MHz internal oscillator)
- In the CONFIG3 register, set WDTE to Off (disable the watchdog)

Then click on the "Generate Source Code" button at the bottom and select all of the generated code (click anywhere in the source code window and use CTRL-A to select all then CTRL-C to copy to the clipboard).

Now in the project directory structure, right click on the Source Files folder and select new main.c, accept the defaults and click on finish.

Paste the clipboard contents into the file below the header and then delete the duplicated #include statement.

Your screen should now look like **Figure 16.5**.

We can now go about the business of setting up PORTA to control our 7-segment display.

Referring to 'Notes on PIC GPIO', to configure a port as output, we know that we need to set:

- TRISx as 0x00 to set all bits as output
- ANSELx as 0x00 to deselect analogue
- ODCONx as 0x00 to disable open drain mode
- WPUPx as 0x00 to disable pull up

Then to write the actual output values we want; we need to write to the LATx register with the value required for the 8 output bits. As we are using 'C' we can do this in decimal, binary or hex. In our case hex is the most convenient as

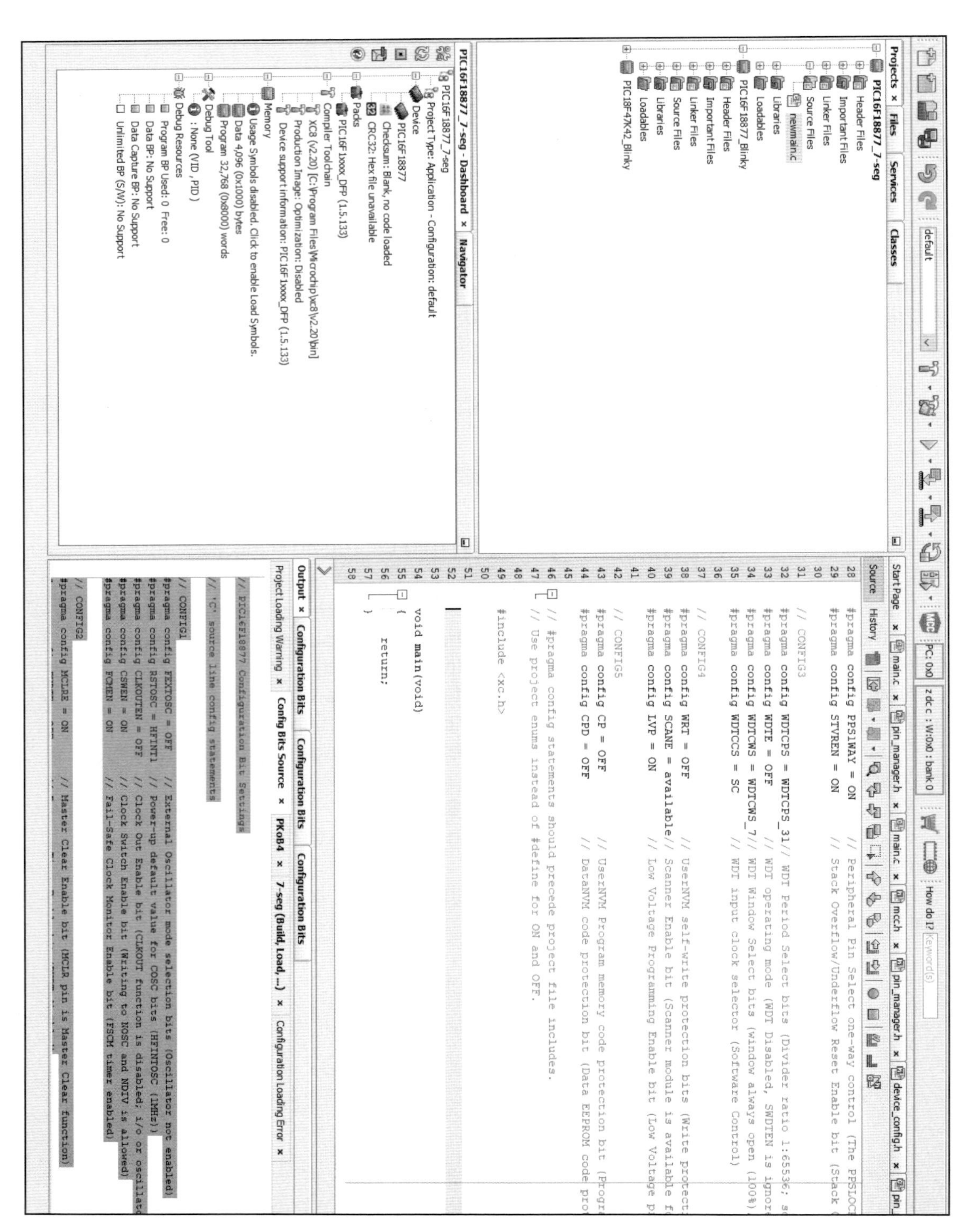

Figure 16.5 - New Source file with config

every 8-bit value can be represented by a 2-digit hex number. As a reminder, you precede the value with '0x' to tell the compiler the value is in hex.

Initially we should set all output bits to 0x00 (for a common cathode) or 0xFF (for a common anode) to switch off all the display elements.

Here is the code we need to do this port initialisation within our main function:

```
TRISA=0x00;         // PORTA is output
ANSELA=0x00;        // clear analogue
ODCONA=0x00;        // disable open drain
WPUA=0x00;          // disable pull ups
LATA=0x00;          // set all outputs to 0
```

We can then introduce a loop within the main function to repeat forever and add the code we want to repeat after the initialisation. If we want to use the __delay_ms() function, we must also add a definition of the oscillator frequency:

```
#define _XTAL_FREQ 1000000
```

above the main function.

Here is an example main function that will repeat the 0 -> 1 -> 2 digits on the 7-segment display changing every 500ms:

```
void main(void)
{
TRISA=0x00;                 // PORTA is output
ANSELA=0x00;                // clear analogue
ODCONA=0x00;                // set to push pull mode

WPUA = 0x00;                // enable pull up

LATA=0x00;                  // set all outputs to 0
while (1)
{
  LATA=0xFC;                // digit 0
__delay_ms(500);
LATA=0x60;                  // digit 1
__delay_ms(500);
LATA=0xDA;                  // digit 2
__delay_ms(500);
  }
}
```

As we did previously, you need to build your project and copy the generated hex file to the flash drive associated with the development board to program the target.

This exercise should have given you an insight into:

- Manually setting up the configuration registers
- Manually configuring GPIO ports
- Writing 8 bits at a time to an output port

Things to Try

- Finish the main function so the code cycles through all the 7-segment digit options including the decimal point
- Do a better job with the code; define an array of the 7-segment display values and use a for loop to select them and place them in the port register in sequence.
- Recreate the project using the Code Generator to set up the configuration registers – decide which method you prefer.
- Add a second 7-segment display to a second port and see if your code can count to 99.

17

Switches and Debouncing

All our GPIO pins can be configured as inputs as well as outputs; we have seen how to use the pins as outputs and now we will experiment with using them as inputs.

From our theory, we know that we need to configure the pins as inputs by setting the associated TRISx register pin to 1 and we will read the input state by accessing the PORTx register.

Debouncing is a common requirement in embedded systems as any mechanical switch is likely to have some bounce associated with the contacts. This is where the contacts physically connect, disconnect, and re-connect as part of a single operation, it might be a very small amount of time that the contacts take to settle, but our MCU can read the input pin status many times during this period, so our input pin status is not guaranteed if the contacts are bouncing.

Debouncing is a technique whereby we read an input pin and then either re-read the pin a short time later to check that the status is the same, or wait a while to avoid reading a bouncing contact.

Hardware Configuration

Let's connect a tactile switch to our development board, one of the pins can be connected to a GPIO pin – I have chosen RD2 – and the other pin connects to ground.

We will also connect a standard LED to a second GPIO pin – I have chosen RD0 – with a current limiting resistor to ground.

We will now write software to light the LED when the tactile switch is pressed.

Software Setup

Open MPLAB X IDE ®, and create a new embedded project for our target PIC using the XC8 compiler and give it an appropriate name.

Once the project has been created, add a new 'C' source file within the project Source Files directory. Use Window -> Target Memory Views -> Configuration Bits to select the internal oscillator at 1MHz and disable the watchdog. Generate the configuration code and paste it into our new 'C' source file.

Define the crystal frequency as 1MHz using:

```
#define _XTAL_FREQ 1000000
```

We now need to configure our GPIO; the switch should be connected to RD2 and the LED RD0.

For the LED:

```
// RD0 as output
TRISDbits.TRISD0 = 0;       // bit 0 of port D
                            // as output
ANSELDbits.ANSD0 = 0;       // switch off analogue
ODCONDbits.ODCD0 = 0;       // switch off open drain
WPUDbits.WPUD0 = 0;         // disable pull up
LATDbits.LATD0 = 0;         // set output to low to
                            // switch off LED
```

And for the tactile switch:

```
// RD2 as input
TRISDbits.TRISD2 = 1;       // bit 2 of port D
                            // as input
ANSELDbits.ANSD2 = 0;       // switch off analogue
WPUDbits.WPUD2 = 1;         // enable pull up
```

Our main loop needs to read the status of the input, set the output as appropriate and delay to avoid any debounce:

```
If (PORTDbits.RD2 == 0)         // if bit 2 of port D is
                                // low (pin RD2)
{
  LATDbits.LATD0 = 1;           // switch on the LED
}
else
{
  LATDbits.LATD0 = 0;           // switch off the LED
}
```

```
__delay_ms(100); // 100ms delay for switch settlement
```

You should now be able to create this project yourself, build and copy the hex file to our target PIC for testing.

A word on Debouncing

Our software above implements a very crude debounce of the input switch contacts using a short 100ms delay between reads of the input pin.

A more correct method is to read the contact status, delay a short while then re-read the contact status to ensure that it is the same and not in a state of change.

There is an excellent debounce algorithm included in the Arduino examples, found under the File option and then->Examples->02.Digital->Debounce.

This Arduino example is well worth an investment of your time to fully understand.

Things to Try

1. Implement the Arduino debounce algorithm in our world of PIC programming.
2. Recreate the project using the MCC – remember to set the analogue selection correctly and include the pull up on the input. These options are found in the resource management (MCC) tab then in the Pin Module.
3. Change the button function so pressing the button toggles the LED status. 1st press = on, 2nd press = off, et cetera.

18

On-Chip Communication Peripherals

Our selected PIC16F18877 contains several on-chip communication peripherals:

- I2C
- SPI
- UART
- CAN

We will look at SPI communications in this experiment with our PIC development board.

SPI

SPI was originally developed by Motorola in the mid-1980s and is designed specifically for short-distance communication in embedded systems. SPI is to be found as the basis of communication between MCUs and many external devices including liquid crystal & OLED displays.

The communications are full-duplex (bi-directional) and synchronised (clocked) with a master-slave architecture, always with a single master and one or many slaves. Each slave device on the bus normally has a separate SS (Slave-Select) or CS (Chip-Select) line which is usually driven low to select communication with the required slave. It is possible to daisy-chain the SS or CS lines, but this makes the software far more complex and this concept is not discussed here.

An SPI bus has 4 specific signal lines:

- SCLK: Serial Clock (output from master)

- MOSI: Master Out Slave In (data output from master)
- MISO: Master In Slave Out (data output from slave)
- CS or SS: Chip Select or Slave Select (usually active low, output from master)

The main advantages of SPI communications are:

- Faster than more basic asynchronous serial communications
- The Receive hardware can be simple
- Multiple slave device support

There are also disadvantages associated with SPI:

- It required 4 wires minimum
- Communications need to be well defined beforehand (you can't just send anything you want to the slave)
- Master is controlling all communications
- Separate SS/CS lines needed for each slave increasing the wire count

The PIC MSSP Module

Our PIC16F18877 device contains a module called the Master Synchronous Serial Port (MSSP) and this is the peripheral that is used for SPI and I²C communications. Our PIC can be a slave, but most commonly would be the master controlling an external or perhaps on-board device.

Chapter 31 of the PIC16F18877 device datasheet provides all the details for the MSSP in our PIC.

The SPI support of the MSSP supports:

- Master
- Slave
- Clock Parity
- Slave Select Synchronisation
- Daisy chaining of the CS or SS lines (this complicates the software requirements)

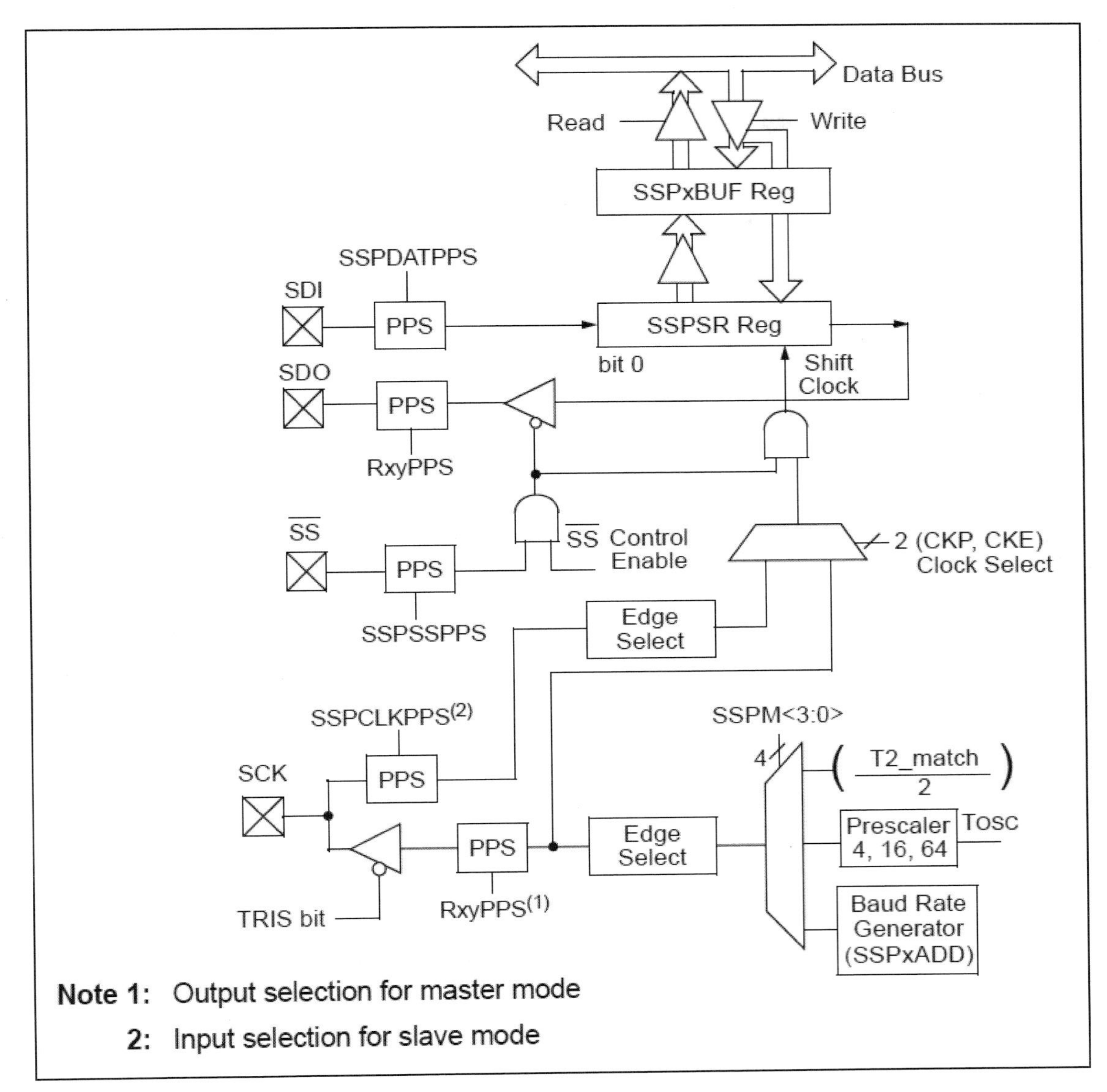

Figure 18.14 - MSSP block diagram - SPI mode

The block diagram of the MSSP is shown in **Figure 18.14** and is reproduced from the device datasheet.

Here we find some potential confusion over naming conventions, the external pins of the PIC MSSP SPI interface are SDI (serial data in – often called MISO), SDO (Serial Data out – often called MOSI), (Slave Select Bar – often called just Slave Select or Chip Select, the bar on the acronym denotes the active low state) and SCK (Serial Clock often also called SCLK).

The PIC16F18877 has 2 SSP (Synchronous Serial Port) modules, SSP1 and SSP2. The CPK and CKE registers of each SSP are used to set the clock polarity and clock edge.

The PIC16F188677 has 5 registers which we will use to configure SPI communications:

- SSPxSTAT - MSSP Status
- SSPxCON1 – MSSP Control Register 1
- SSPxCON2 – MSSP Control Register 2
- SSPxBUF – MSSP Buffer Register
- SSPxADD – MSSP Address Register
- SSPxSR – MSSP Shift Register (not directly accessible)

It is noteworthy that when using an SPI communication, we still need to separately configure the pins being used as input or output appropriately.

The details of each of the registers above can be found in the device datasheet, but always remember that the MSSP can support different communication protocols (SPI and I²C) and therefore some of the configuration bits are not used in specific modes.

For example, the SSPxSTAT (MSSP Status Register) only has 3 bits that are used when configured as SPI. Bit 7 controls the data sample point, Bit 6 selects the required clock edge and bit 0 shows when the receive buffer is full. The other 5 bits of the SSPxSTAT register are not used in SPI mode.

An ADC based Voltmeter

The MCP3021 is a low-cost ADC device which provides data communication over SPI. We will interface this device to our development board and write software to retrieve the data and display it on an LCD display.

This exercise is included as an example of SPI communications using the PIC registers.

The interface to the LCD is provided by a simple library routine that can be found in the RSGB book repository in directory: PIC\LCD Library

These files should be included within the MPLAB project as appropriate (.c file in source files and .h file in header files).

To set up our SPI interface, we first need to declare the pins we are going to use as inputs and outputs:

- RA5 – output SPI Slave Select or Chip Select (SS or CS)

- RC3 – output SPI Clock (CLK or SCLK)
- RC2 – input SPI Data In (SDI or MISO)
- RC4 – output SPI Data Out (SDO or MOSI)

Then, in accordance with the datasheet, we need to switch off interrupts and unlock the PPS (Peripheral Pin Select) module to allocate the selected GPIO pins to the MSSP. To unlock the PPS, the datasheet says we must:

- Write 0x55 to the PPS Lock Register (PPSLOCK)
- Write 0xAA to the PPS Lock Register (PPSLOCK)
- Set the PPSLOCK bit of the PPSLOCK register to 0x00 to unlock the PPS for configuration

To allocate our chosen GPIO pins we would use:

- SSP1DATPPSbits.SSP1DATPPS=0x12; //set SSP1 SDI to be RC2
- RC3PPS=0x14; // set SPI CLK to be RC3
- RC4PPS=0x15; // set SPI SDO to be RC4 (not used in MCP3201)

Note: the SS or CS pin is simply used as a GPIO line to enable the slave, there is no mapping needed to SSP.

Then we lock the PPS:

- Write 0x55 to the PPS Lock Register (PPSLOCK)
- Write 0xAA to the PPS Lock Register (PPSLOCK)
- Set the PPSLOCK bit of the PPSLOCK register to 0x01 to lock the PPS

Finally, to initialise our SPI communications, we need to:

- Set SSP1STATbits.SMP=1 // 1 = Input data sampled at middle of data output time
- SSP1STATbits.CKE=1 // 1 = Transmit occurs on transition from active to Idle

- SSP1CON1bits.SSPEN=1 // 1 = enable the serial port
- SSP1CON1bits.CKP=0 // 0 = normal clock polarity
- SSP1CON1bits.SSPM=10 // SPI Master mode, clock = FOSC/(4 * (SSPxADD+1))(5)
- SSP1ADD=0xFF // 31.25kHz clock

Note: clock frequency calculation is defined by equation 31-1 of the device datasheet.

We are now ready to read from the SPI bus the data that is being sent by our external ADC device.

The datasheet for the MCP3021 shows us yet more complexity, the data sent from the device is sent in two separate bytes, we then need to extract the lower 4 bits of the Upper Data byte and the upper 6 bits from the Lower Data Byte to construct the 10-bit ADC value.

We can read from our SPI device and do the necessary jiggery-pokery to the data bytes as follows, first to read in the 2 bytes of data from the SPI:

```
LATAbits.LATA5=0;              // enable CS active low
msb = SSP1BUF;                 // dummy read to clear BF
SSP1BUF = 0x55;                // dummy write to provide clock
                               // for read operation
while(!SSP1STATbits.BF);       // wait for the reception of data
        msb=SSP1BUF;           // read in the byte received
                               // (Upper Data Byte)
        SSP1BUF=0x55;          // dummy write to provide clock
                               // for read operation
while(!SSP1STATbits.BF);       // wait for the reception of data
        lsb=SSP1BUF;           // read in the byte received
                               // (Lower Data byte)
LATAbits.LATA5=1;              // disable CS
```

Now to extract the 10-bits we need:

```
ADCValue=(msb & 0x1F);         // but mask lower 4 bits
ADCValue = ADCValue <<7;       // move the result 7 bits left
```

```
lsb=lsb>>1;                          // move the upper 6 bits of the
                                     // lower byte right 1
ADCValue = ADCValue | lsb;           // join the two values with an 'or'
```

Finally, we can convert our data value to a voltage with:

```
Voltage= ADCValue*5.0/4096;          // 5V is our device VDD
```

The complete code to create our voltmeter experiment is contained in the RSGB book file repository under the directory PIC.

Building the Project

Our 4-bit interface to the LCD module is wired as follows:

LCD Pin	MCU Pin
RS	RE0
E	RE1
DB7	RD7
DB6	RD6
DB5	RD5
DB4	RD4

We should already know the MCP3021 wiring from the code we have written above.

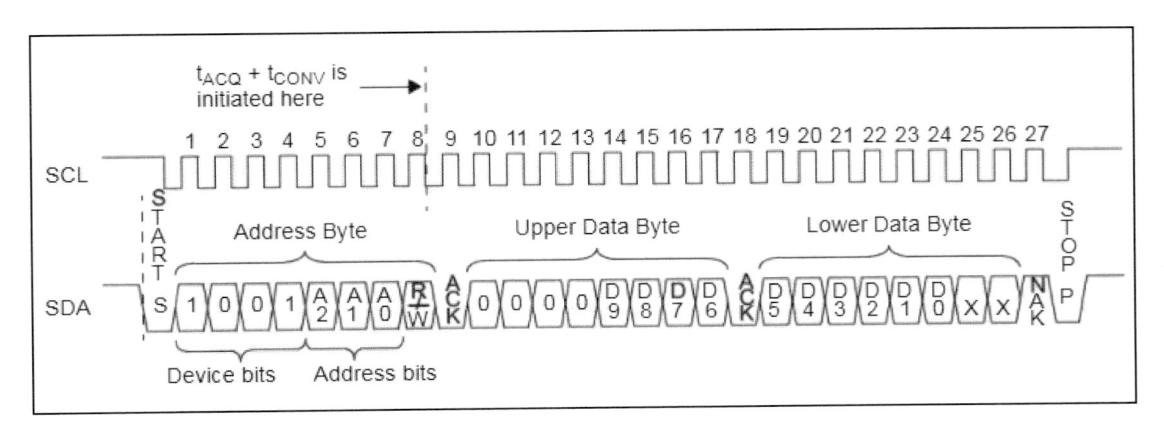

Figure 18.15 MCP 3021 data send.

Once you have set up your hardware, build the project in the MPLAB X IDE ® and copy the generated hex file to the development board flash drive.
You should find you now have a fully accurate voltmeter reading the voltage applied to the input of our MCP3021 device.

Things to Try

1. Connect the input of our ADC to a potentiometer so you can vary the device input voltage from VDD to 0.
2. Replicate the project using an alternative device like the ADS1115 ADC which uses I^2C communication.

19

Notes on PIC Debugging

When we used the STM32CubeIDE we learned that one of the biggest advantages offered by this development environment was the ability to debug our code on the target. This was an extremely powerful tool which was not available on the more simplistic Arduino IDE.

The MPLABÆ IDE does support similar debugging capabilities to the STM-32CubeIDE; however, this is not available on our low-cost development board.

If you decide to explore PIC programming in more depth, then either the MPLAB® PICKit in-circuit debugger or a more advance development board would be a good investment as this will enable the on-target debugging facilities available in the MPLAB® IDE.

The picture **Figure 19.1** shows the authors PICKit in-circuit debugger and a Curiosity High Pin Count development board – both support the MPLAB® IDE debugger.

PIC to PC Serial Communications

As we learned with our Arduino coding examples, very basic debug information can be communicated from our target device back to the PC using serial communications. This enables the developer to embed statements throughout the code under test to pass information back to the PC screen as the code executes. We introduced the concept of a Boolean switch variable to turn debug information on and off in our examples and here we will explore how to send serial data back to our host PC from our low-cost development board.

To facilitate the PC communications, we will use the same FTDI USB to Serial adaptor that we used with our STM32 boards.

Hardware Setup

Table 13-1 of our PIC16F18877 datasheet shows that the default pins for the on chip UART peripheral are RC7 for RX and RC6 for TX.

Figure 19.1 - More advanced PIC debug support

We therefore need to wire the RX line of our FTDI USB to Serial adaptor to pin RC6 and the TX line to RC7. As was previously we need to ensure that the FTDI USB to Serial adaptor is set to operate in 3V3 mode.

Software Development

Open MPLAB X-IDE® and create a new project for our target device.

Launch the MPLAB® Code Configurator and under the Easy Setup tab make sure we have the internal oscillator enabled and the watchdog disabled.

Under the device resources find the EUSART section and expand it, then add the EUSART to the project by clicking on the green plus sign as shown in **Figure 19.2**. (see following pages)

As we are using this for one-way PIC to PC comms we can enable the EUSART and Transmit, also select the option to redirect STDIO to the EUSART as shown in **Figure 19.3**. (see following pages)

Click the Generate button to create the MCC generated files and then open the main.c file created for us.

We need to include the MCC generated EUSART code by adding the line:

#include “mcc_generated_files/eusart.h”

Then to test our communications, we can make our main loop look like this:

```
while (1)
{
    // Add your application code
    putch('X');
    __delay_ms(200);
}
```

Build the project and copy the hex file to the board flash drive device to programme our target.

Open and configure your favourite serial terminal program on your PC and you should now see a string of Xs appearing down the serial port as shown in **Figure 19.4**.

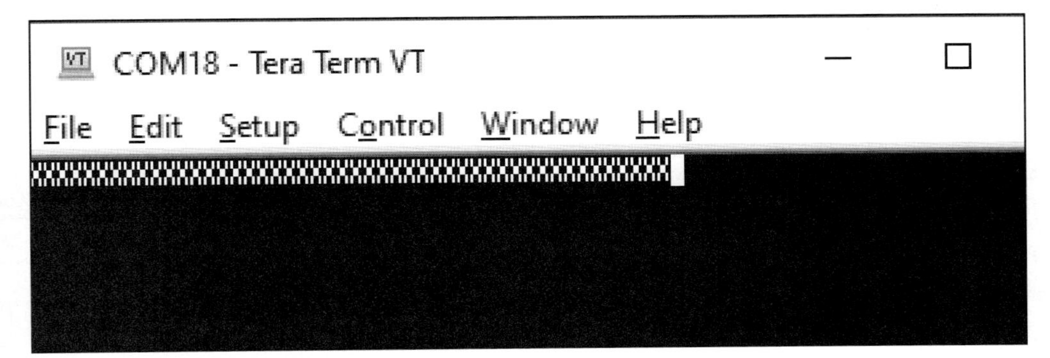

Figure 19.4 - Serial Comms from the PIC to PC

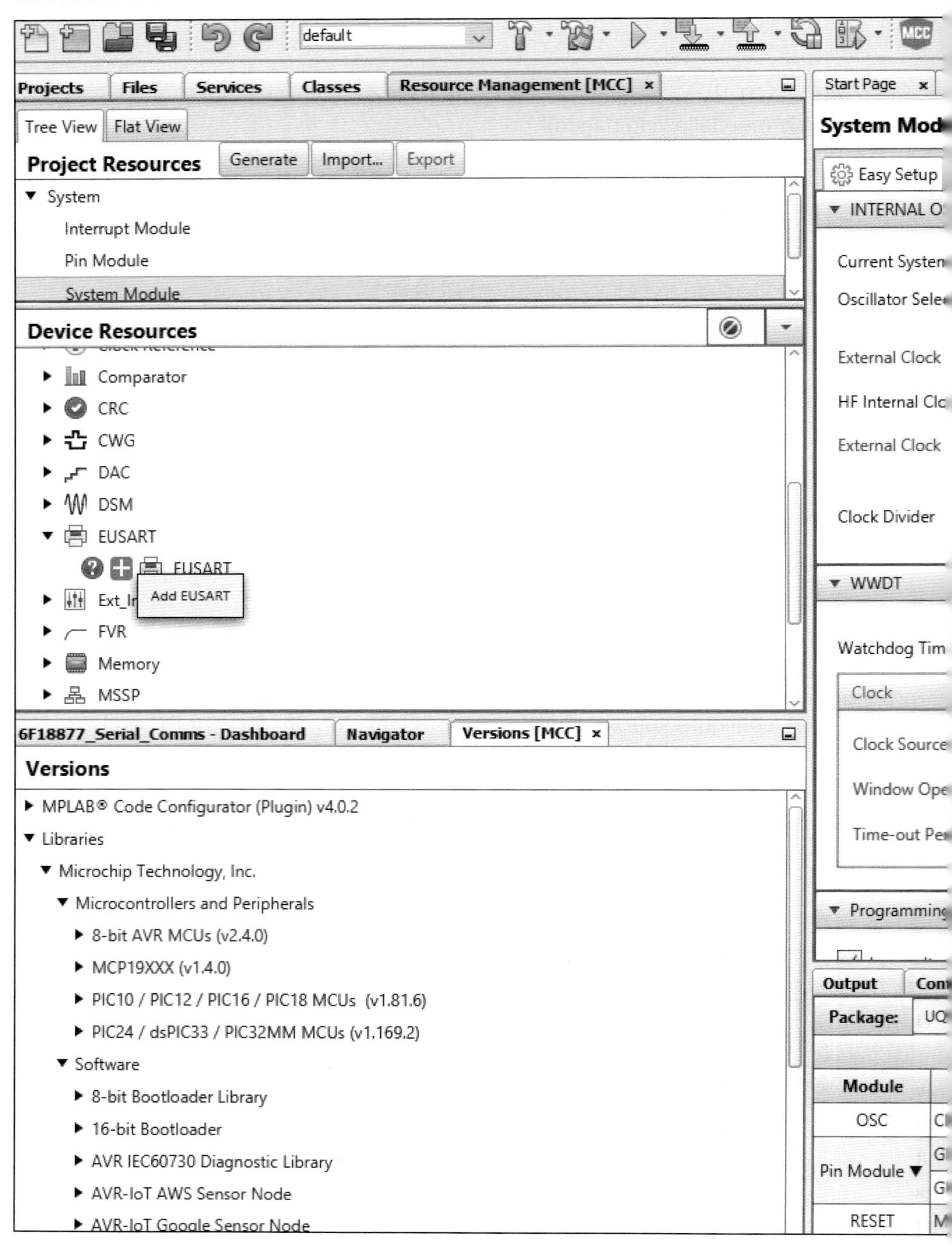

Figure 19.2 - Adding the EUSART resource

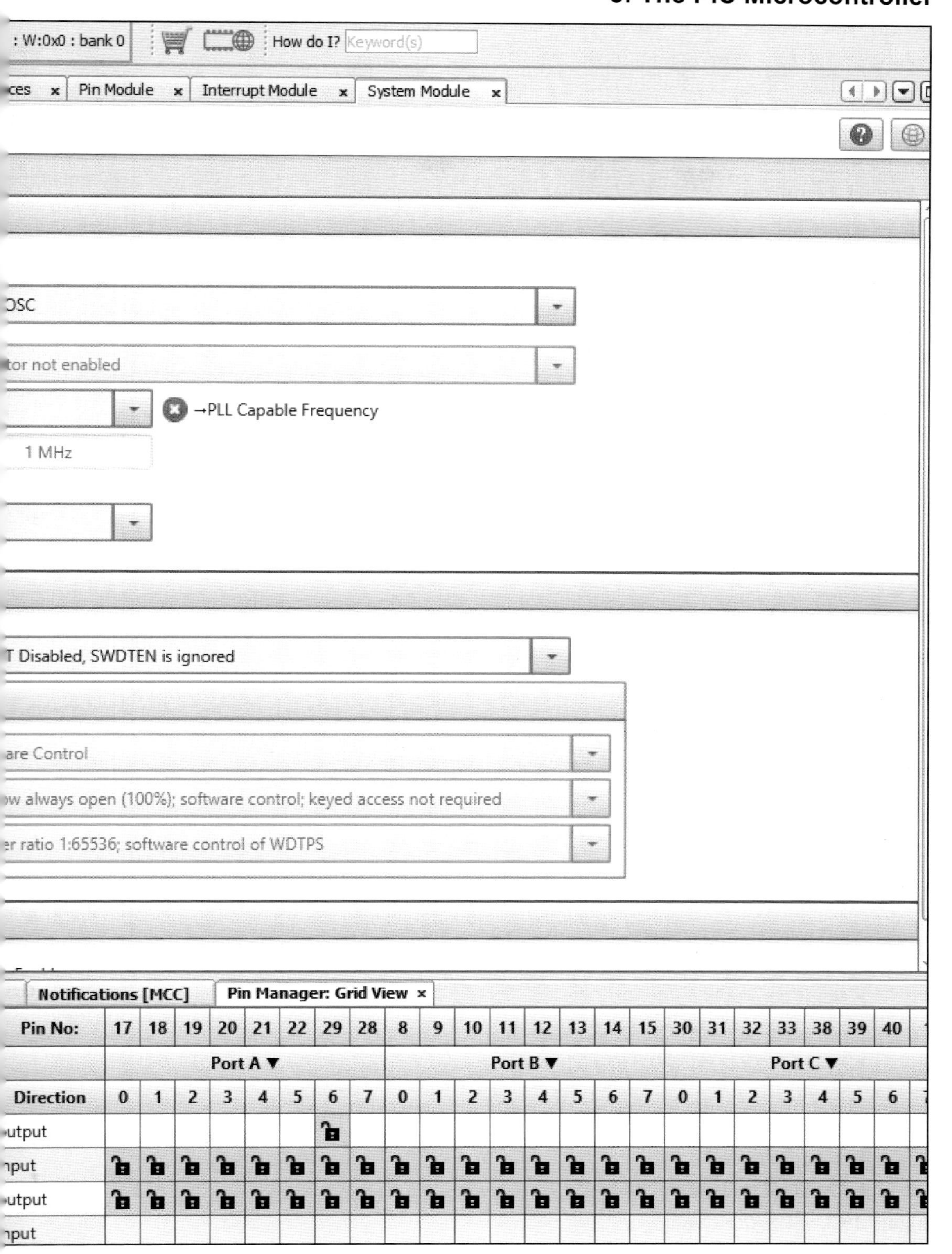
: W:0x0 : bank 0
How do I? Keyword(s)
ces
Pin Module
Interrupt Module
System Module
OSC
tor not enabled
→PLL Capable Frequency
1 MHz
T Disabled, SWDTEN is ignored
are Control
ow always open (100%); software control; keyed access not required
er ratio 1:65536; software control of WDTPS
Notifications [MCC]
Pin Manager: Grid View
Pin No: 17 18 19 20 21 22 29 28 8 9 10 11 12 13 14 15 30 31 32 33 38 39 40
Port A
Port B
Port C
Direction 0 1 2 3 4 5 6 7 0 1 2 3 4 5 6 7 0 1 2 3 4 5 6
utput
nput
utput
nput

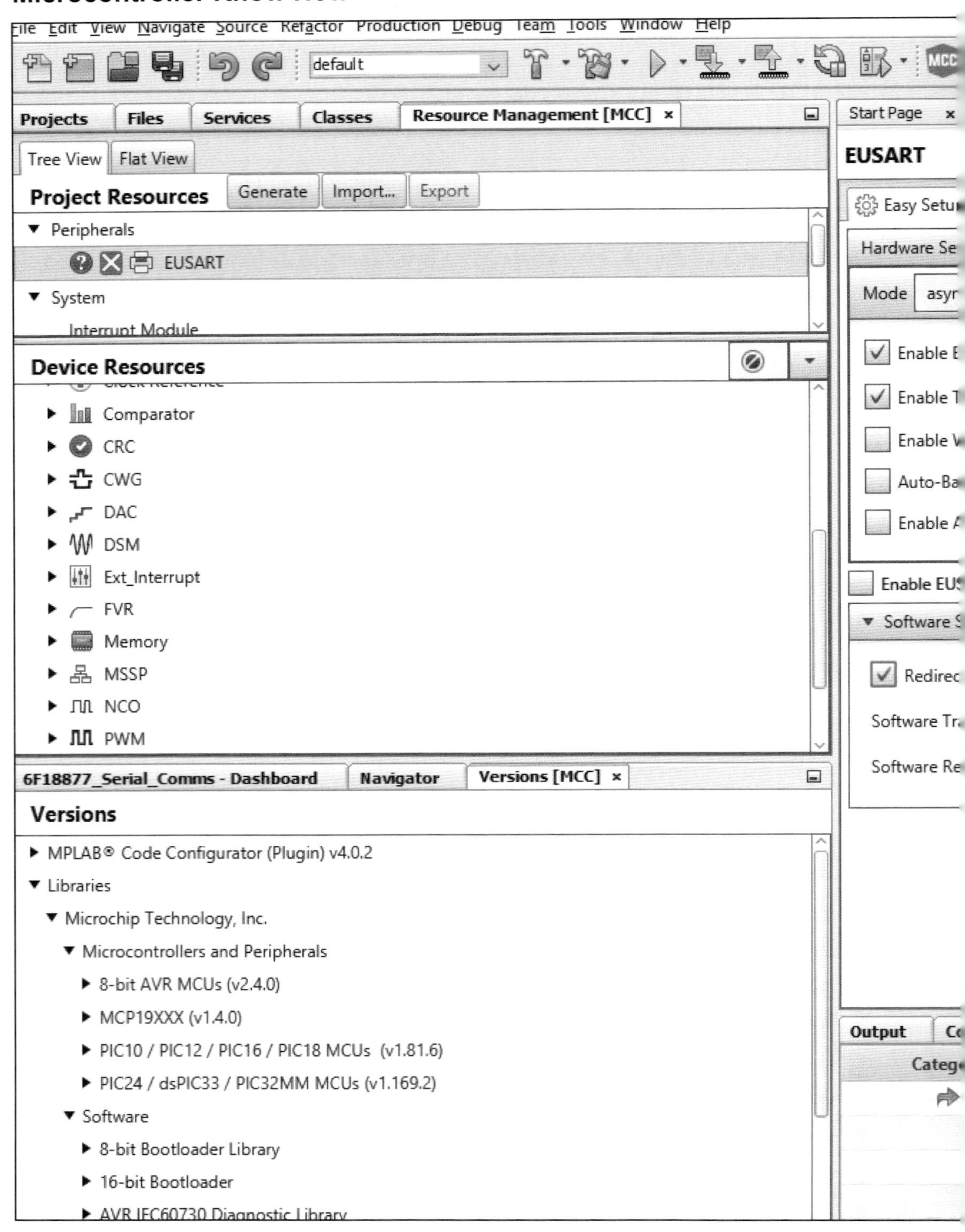

Figure 19.3 - EUSART Config

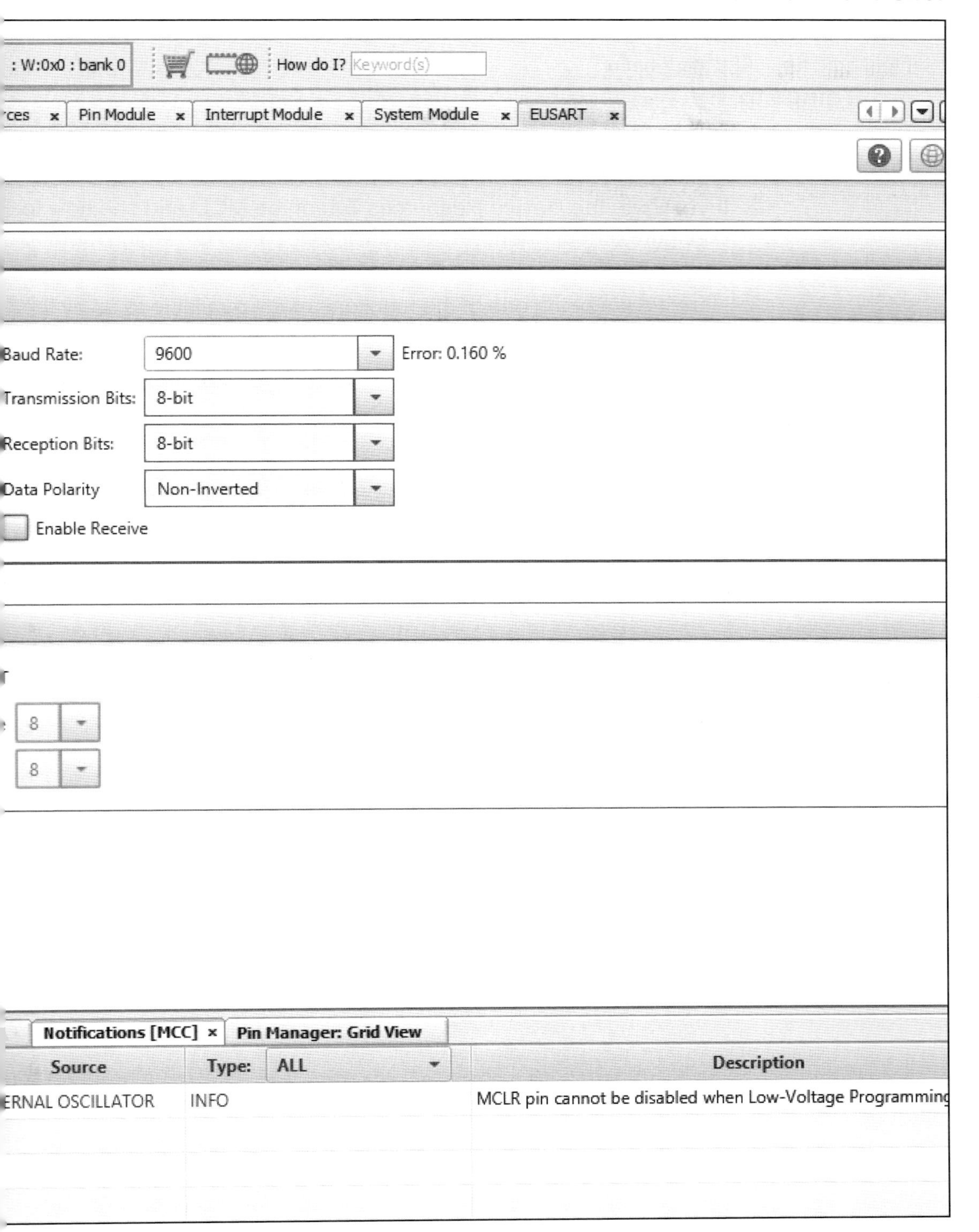
: W:0x0 : bank 0
How do I? Keyword(s)
rces
Pin Module
Interrupt Module
System Module
EUSART
Baud Rate: 9600
Error: 0.160 %
Transmission Bits: 8-bit
Reception Bits: 8-bit
Data Polarity Non-Inverted
Enable Receive
8
8
Notifications [MCC]
Pin Manager: Grid View
Source
Type: ALL
Description
ERNAL OSCILLATOR
INFO
MCLR pin cannot be disabled when Low-Voltage Programming

Things to Try

1. During our STM32 experiments we developed a nifty PrintSerial routine; modify this routine for use in our PIC experiments.
2. Investigate the Device Resources Tab of the MCC and see how to configure other peripheral devices like SPI and I^2C – which method do you prefer?
3. Here we are using one of the functions made available to us by the MCC 'putch'; investigate the full function set within the eusart.h file.

20

Additional Musings

Monitoring code execution speed

The Arduino IDE gives us a neat function called millis(). This useful embedded function returns the number of milliseconds passed since the MCU began running the current program. This number will overflow (go back to zero), after approximately 50 days and returns an unsigned long variable.

In the STM32 Cube IDE we have a similar function provided by HAL called HAL_GetTick().

We can use these functions to determine the time (in milliseconds) that our code is taking to execute. This can be useful if you think a particular block of code is causing problems, or if you want to see just how slick your coding skills are.

Here is a rather silly piece of iterative code to (sort of) calculate Pi. The MCU must include a floating-point co-processor as this will be utilised in the calculations:

```
StartTime = millis();
for (float k = 0.0; k <= numofTerms; k++) {
  pi += 4.0 * ( pow((-1.0), k) ) * ( 1.0 / (2.0 * k +
1) );
}
EndTime = millis();
```

Remember that if the EndTime is < StartTime the timer has wrapped round and needs to be adjusted accordingly.

Defining the loop control variable numofTerms as 100,000 the following timings (in seconds) for the loop execution have been noted on different MCU boards:

30.1499996185302734375O	Arduino Mega
29.6009998321533203125O	Arduino Uno/Nano
7.46299982070922851562	Arduino Zero
7.45300006866455078125	SAMD21 Mini
3.29900002479553222656	STM32F303
3.07200002670288085937	STM32F103C (Blue Pill)
2.96900010108947753906	Arduino Due
2.89599990844726562500	STM32L432KCU6
1.87699997425079345703	STM32F411
0.66799998283386230468	ESP8266

Sending Accurate CW

To send accurate CW from a microcontroller, first we must understand the definition of the language that is our Morse Code.

Firstly, we need to define the speed of our sent code in WPM (Words per Minute). In a standard Arduino sketch we can do this as:

```
int wpm = 15;          // Words per minute CW speed
```

Secondly, we can then define the length of a dit (or the dot if you prefer) as:

```
float ditf = 1200/wpm; // dit length in ms is 1200/WPM
```

The value above is a real number (contains decimals) and is defined as floating point.

Next, we can convert the floating point value to the nearest integer (useful if we want to use the delay function):

```
int dit = ditf;        // convert the floating point maths
                       // to an integer so we can use the
                       // delay function
```

Now we can define the dah (or the dash if you prefer) as:

```
int dah = 3 * dit;     // dah is 3 dits
```

Finally, we can define the spaces we need between letters, characters and words:

```
int lspace = dah;      // letter space is the same as dah
int cspace = dit;      // character space is the same
                       // as a dit
```

```
int wspace = 7 * dit; // word space is 7 dits
```

Using these definitions, we can very easily create a beacon using the CW TX project as a starting point.

We can create a set of routines that will provide us with a dit, dah, character, word and letter spaces:

```
void dot ()              // key down for the dot duration
{
  digitalWrite(cw_key_line, LOW);
  delay(dit);
  digitalWrite(cw_key_line, HIGH);
  char_space();
}

void dash ()             // key down for the dash duration
{
  digitalWrite(cw_key_line, LOW);
  delay(dah);
  digitalWrite(cw_key_line, HIGH);
  char_space();
}

void word_space ()       // word space
{
  delay (wspace);
}

void letter_space ()     // letter space
{
  delay(lspace);
}

void char_space()        // character space
{
  delay (cspace);
}
```

Then we can define individual characters and words:

```
void char_g ()
{
  dash();
  dash();
  dot();
  letter_space();
}

void char_m ()
{
  dash();
  dash();
  letter_space();
}

void char_x ()
{
  dash();
  dot();
  dot();
  dash();
  letter_space();
}

void char_0 ()      // you will get the idea by now.....
{
  dash();
  dash();
  dash();
  dash();
  dash();
  letter_space();
}


void g0mgx ()
{
  char_g();
  char_0();
  char_m();
  char_g();
  char_x();
}
```

Taming the COM Port

If you connect multiple boards to your PC over time, especially if you use multiple different COM ports, you will observe the COM port number assignment of your board increasing rather alarmingly. You may soon find yourself in quite a muddle with potentially hundreds of COM ports defined on the Windows PC.

There is a way to tame this scenario, but not one that seems to be well documented.

Here is an example of the muddle I got into recently:

Each of the ports listed in light grey is a definition that isn't currently connected. Under normal use of the Device Manager, these devices are hidden.

However, if we execute the following command from a Command Prompt on the PC:

```
set devmgr_show_nonpresent_devices=1
```

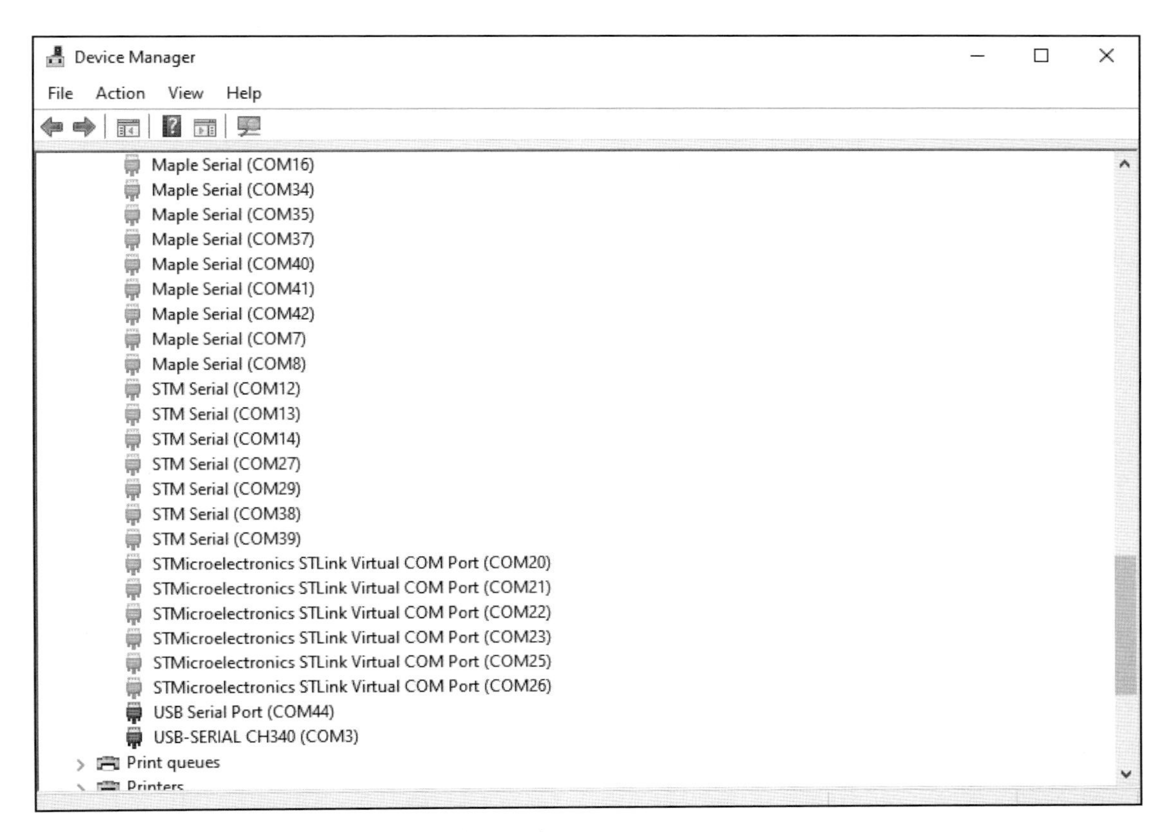

Figure 20.1 - COM port numbers out of control

then start Device Manager using:

devmgmt.msc

When from the view menu in the Device Manager I select "Show Hidden Devices" all of your unused COM ports in the PORTS (COM & LPT) section are shown.

Once visible, you can select a COM port by clicking on it and press the delete key – voila! Your COM port number is available again.

Performing this task from time to time can keep your COM ports under control and stop the numbers escalating out of control.

21

Conclusions

Through the experiments and projects in this book I have tried to introduce you to a number of different low cost MCUs, we have learnt to use a number of different IDEs by different manufacturers and have looked at the low-level configuration needed to get MCUs and their peripherals to function.

If this world of embedded programming interests you, I expect this to be the start of a very long journey; the world of embedded software is now very accessible to the hobbyist with vast resources available on-line and free.

You should probably focus on one manufacturers devices and toolset, avoiding the trap I have fallen into of being 'jack of all trades and master of none'.

But whatever you do with embedded programming and MCUs, remember to have fun!

73s Mark, G0MGX

RSGB BOOKSHOP

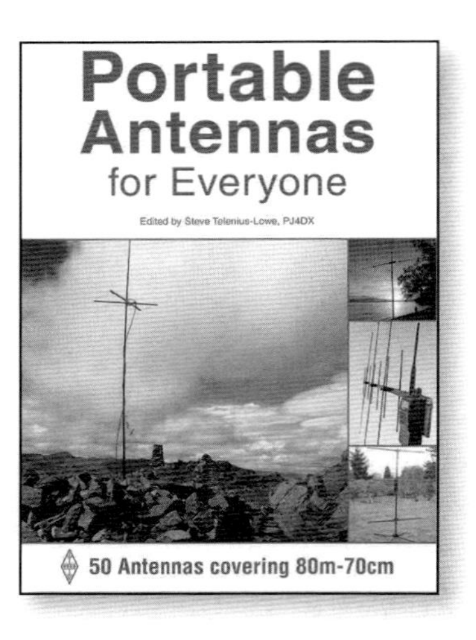

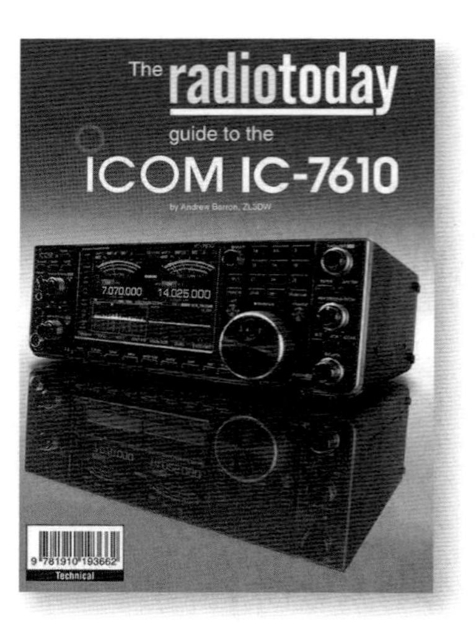